I0045071

Edward T. (Edward Travers) Dixon

A Paper on the Foundations of Projective Geometry

(Read before the Aristotelian Society, Dec. 13, 1897)

Edward T. (Edward Travers) Dixon

A Paper on the Foundations of Projective Geometry
(Read before the Aristotelian Society, Dec. 13, 1897)

ISBN/EAN: 9783337008833

Printed in Europe, USA, Canada, Australia, Japan

Cover: Foto ©berggeist007 / pixelio.de

More available books at **www.hansebooks.com**

A PAPER

FOUNDATIONS

OF

PROJECTIVE GEOMETRY

(Read before the Aristotelian Society, Dec. 13, 1897)

BY

EDWARD T. DIXON

PREFACE.

This paper was nominally "read" at a meeting of the Aristotelian Society on December 13th, 1897. It is the excellent custom of that Society to circulate proofs of papers among the members before the meetings, and I was thus able to make the paper longer and more technical than would otherwise have been possible; but even so the mathematical portion is compressed to the utmost, and a great deal is necessarily left to the good will and intelligence of the reader.

Since writing the paper I have read up the modern literature of the subject, and I am confirmed in the impression, which I got from Mr. Russell's book, that the conception of "order" as the basis of so-called "Projective Geometry" is essentially new. The only suggestion of this way of looking at the question which I have found is in Pasch's *Vorlesungen über neuere Geometrie*. But though he sees that less than four points have no determinate "order," he never extends the conception of order to continuous quantities, nor does he grasp the connection between it and "anharmonic ratio." It seems, therefore, worth while to bring my views before a wider public, for, interesting as the discussion before the Aristotelian Society was, it was chiefly my epistemological views which came in for criticism; the geometrical points were hardly discussed at all.

EDWARD T. DIXON.

January, 1898.

THE

FOUNDATIONS OF PROJECTIVE GEOMETRY

PART I

I HAD the honour of reading a paper on Natural Realism before this Society last year, the object of which was not so much to convert Idealists to realism (for I held, and hold still, that they are all natural realists except in moments of philosophic extasy), but rather to discover if possible what modern Idealism really is. I failed; perhaps because my remarks were too general in character. My opponents in argument, while professing to be Idealists, adopted my conclusions and gave them their own interpretation, so that apparently nothing was left to argue about. I propose therefore to-night to discuss a special case only, our so-called knowledge of space, and to see whether it does not provide a test which shall discriminate between the methods, even if not finally between the opinions, of idealistic and realistic philosophers. I shall do so with special reference to a book which has recently appeared on the subject by Mr. Russell, a member of this Society, who is himself, if I mistake not, an Idealist of the modern type.

Mr. Russell commences his book with the statement that throughout the 17th and 18th centuries Geometry remained, in the war against empiricism, "an impregnable fortress of the idealists," meaning that they successfully maintained the position that in Geometry at least we had obtained certain knowledge about the real world, independently of experience. "None but a madman, they said, would throw doubt on its validity, and none but a fool would deny its objective reference." He implies that modern idealists have receded from this position; and it is the primary object of this paper to discover whether they really have done so or not. Mr. Russell's book professes to be an exposition of the position modern idealists take up in reference to this question. We shall see whether it differs in any essential from that of the idealists of the last two centuries.

Before proceeding to consider in detail the constructive portions of Mr. Russell's book, I must notice a point in which his critical method, and that of the school of philosophy to which he belongs, seems to me radically defective; and this is the treatment of contradictions and 'antinomies.' It is true that all Mr. Russell does is to attempt, and generally with success, to explain them away. But this is not the case with all modern philosophers; and that Mr. Russell allows himself ever to attach any importance to so-called 'antinomies,' without at once holding them up to ridicule, shows how much he, like other modern philosophers, is still under the pernicious influence of that arch-sophist Immanuel Kant. Consider for example the 'antinomy of the point'—"There are different points: all points are identical." Surely it is evident at once that this is only a clumsy way of saying, "There are points which differ in position, but are in other respects identical." Mr. Russell's own statement of the antinomy is expressed in rather more specious, or at least more involved, language; but I think he might have explained it away in half the space he actually devotes to it. In his treatment of the second antinomy however he is not so happy. As stated by him, the contradiction arises solely from an ambiguous use of the term 'relation'; and so far from explaining this ambiguity away, it remains, and mars much of the reasoning throughout his book. Surely he ought to have seen that when he says spatial figures (he is referring to straight lines and planes) must be regarded as 'relations,' he is using the word in a completely different sense to that which it bears in the assertion that 'a relation is indivisible.' To me it has always seemed that to talk of a straight line as a 'relation,' or still more as 'the' relation between two points, is an abuse of language. One might indeed fairly say that the fact that one, and only one, straight

line can be drawn through two points indicates the existence
of a relation between them; but the relation is there whether
you draw the line or not; it is certainly not the same thing as
the line. But if it is desired, or found convenient, to call a
straight line 'the relation' between two points it is of course
logically permissible to do so, *provided* that some other term,
such as 'relationship,' is used for the kind of abstract conception which can not be divided.

But it is simple waste of time to discuss so-called antinomies in detail. To talk of a contradiction in things,
whether objective or subjective, is sheer nonsense. To do so is
in itself simply to make a contradiction in terms. If you
attempt to apply two contradictory terms, or one which has
been defined in a self-contradictory manner, to any real thing,
it simply won't fit. You can not by doing so affect the nature
of the thing, nor find out anything, either positively or
negatively, about it. If your language is self-consistent and
free from contradictions you may employ its terms as symbols
for real things; and in this way it may assist you to obtain
clear conceptions with regard to them. But if it is not self-
consistent, it is mere 'gibberish'; and it passes my compre-
hension to understand how anybody could imagine that terms
which were admittedly contradictory could be of any use. If
anyone wishes to see the absurdities to which this sort of
'logic' can lead, let him read Mr. Bradley's "Appearance and
Reality."

Mr. Russell devotes a considerable amount of space in his
introduction to the defining of the terms '*a priori*' and
'wholly *a priori*.' I shall show that there is some reason to
doubt whether he really does confine himself to the definitions
of the terms which he gives at the beginning of his book,
when he uses them in the statement of his final conclusions.
But at present I wish to discuss the propriety of using the term
a priori at all. The term is philologically and historically
bound up with the view "that certain knowledge independent
of experience is possible about the real world," and if you do
not want to express this view it is better not to use a word
which in the minds of the enormous majority of your readers

will certainly imply it.   However, Mr. Russell defines the term
a priori in a new and special sense, in which *a priori*
knowledge, even if 'certain,' and 'independent of experience,'
may perhaps not be intended to be knowledge of the real
world.   Whether he so intends it or not I am not able with
certainty to determine from what he says in his introduction ;
for he declines to discuss any psychological question.   Even if
he agreed with Kant, that *a priori* knowledge is subjective,
this would not settle the question, for he gives two discrepant
definitions of 'subjective,' one of which, without any apparent
justification, he fathers on Psychology, and the other on
Physical Science.   Nor is it any clearer what he means by
'experience.'   He refers to Mr. Bradley for his definition, who
certainly includes sensations in the denotation of the term *
But Mr. Russell can not seriously mean that he believes in the
a priori character of the axioms of projective geometry
because without them a toothache would be impossible !   If he
had said those axioms were only *relatively a priori*, as he says
of the axioms of metrical geometry, he would only have
meant, according to his definition, that a science deduced from
other axioms would not be the science of projective geometry—
which is only another way of saying that the axioms do in fact
define the terms of the science—they are logically merely
definitions.   With that I entirely agree, but what more does
Mr. Russell mean by 'wholly' *a priori* ?   That they are true of
any form of externality ?   But is not this the same thing over
again ?—if I admit this, does it not simply define the term
'externality' ?   I might have some other theory of reality, not
based on Mr. Russell's axioms, and all he has proved against
such a theory would be that I had no right to call it a theory
of externality.   I have pointed out elsewhere† that, unless we
define our terms independently but not inconsistently, no
proposition can be held to be more than an implicit definition
of any doubtful term it may contain.   It was of course open
to Mr. Russell to have defined the term 'externality' indepen-

* See "Appearance and Reality" p. 346.
   † See my *Essay on Reasoning*, Deighton, Bell & Co.   Also an article in
Mind, N. S. Vol. II., p. 339.

dently and consistently. He might for example have defined it as the conception we actually have, commonly called Space. This would be what I have called a definition by denotation; and the other terms being defined by connotation, the assertion, that certain axioms must be true of any form of externality, would acquire real import. But this would have involved Mr. Russell in a psychological discussion as to how we actually do conceive space, and would have made the axioms dependent upon subjective experience. What Mr. Russell is trying to do, and it is herein that his idealism appears, is to obtain knowledge which shall be real, and yet not be "at the mercy of empirical psychology." He wants to know something about the psychological world, if not about the physical, '*a priori*'; in the good, old, undiluted sense. We will see how far he succeeds.

He commences the detailed discussion of the elements of projective geometry by saying (§ 111) "The two mathematically fundamental things in projective geometry are anharmonic ratio and the quadrilateral construction." These things may indeed be essential to the mathematical treatment of the subject, but there are a number of preliminary propositions which must be established before we can discuss them, which Mr. Russell tells us nothing about at all, and he hardly tells us anything even about the 'axioms' upon which he conceives them to be based. He does indeed enumerate these axioms in a subsequent paragraph (§ 122); and I suppose he relies on his readers' knowledge of the ordinary text books to effect the requisite deductions. But in none of the ordinary text books, so far as I am aware, are the preliminary propositions deduced from axioms at all like Mr. Russell's, or even from definitions which profess to be free from spatial implications. The one possible exception with which I am personally acquainted is Veronese's important work on the elements of geometry; but though Mr. Russell quotes from this work he can not have appreciated its importance, or he would have given more attention to those preliminary propositions about which he practically tells us nothing at all. For example, in the diagram on p. 123 of Mr. Russell's book, if the section *abcd* is

given, and two points a'' c'' on the third section, how does Mr.
Russell deduce from his axioms that there is a point b'' at all?
And how that there is a point d'' at all?

Mr. Russell's treatment of 'anharmonic ratio' and the
'quadrilateral construction' is just enough to whet the appetite
for knowledge without satisfying it. He tells us that two sets
of four collinear points each have the same 'anharmonic ratio'
when corresponding points of different sets lie two by two on
four straight lines through a single point, or when both sets
have this relation to a third set. He adds, in a note, "There is
no corresponding property of three points in a line, because
they can be projectively transformed into any other three points
on the same line." But if this is so he expresses himself very
badly when he tells us that "The successive application to any
figure of two reciprocal operations of projection and section is
regarded as producing a figure projectively indistinguishable
from the first." He does indeed make his meaning clearer later
on, by saying that "Two sets of points or lines *which have the
same anharmonic ratio* are treated by projective geometry as
equivalent," (my italics) that is, two sets of *four* points obtained
by projection and section are regarded as indistinguishable; for
we know there is no anharmonic ratio between three. But
what is meant by the term 'anharmonic ratio' if it is not used
solely on account of its numerical, if not metrical, significance?
In Mr. Russell's exposition it seems merely designed to lead up
to the application of analysis to the problems of projective
geometry, by means of the quadrilateral construction. Why,
for example, does Mr. Russell assign certain numbers to the
points A, B, C, D in his quadrilateral construction? In order,
he says, that the differences AB, AC, AD may be in *harmonical*
progression. Is this merely because we called it an *anharmonic*
ratio? Or is it *merely* in order to fit in with preconceived
spatial notions? If Mr. Russell's preconceived notions had
been those of, say, spherical geometry, he would not have made
the differences a harmonical progression. Does the distinction
between the meanings of anharmonic ratio in Euclidian and
spherical geometry depend upon nothing of which we can take
cognizance without the notion of number? Or if we did not

care to apply numerical analysis to points at all, would anharmonic ratio be devoid of all significance ?

To answer this we must attempt to realise what significance projective geometry could have, apart not only from spatial, but even from numerical applications. Mr. Russell does attempt an answer in §§ 117—124 in his book. He commences by giving a definition which he admits contains a 'fundamental contradiction'; but, not in the least deterred, he goes on to define 'quality in geometrical matters' in the belief that the rest of his definitions will 'follow without difficulty'; and in the hope that they will enable him to discover 'what sets of figures are projectively indiscernible.' This he conceives himself to have done in § 119, where he tells us that points " have no intrinsic properties; but they are distinguished solely by means of their relations. Now the relation between two points is the straight line on which they lie. This gives that identity of quality for all pairs of points on the same straight line which is required by our projective principle. If only two points are given they can not without the use of quantity be distinguished from any other two points on the same straight line." Very well; but unfortunately, Mr. Russell sets out to prove this of *three* points in a line, not of two, and it does *not* take three points to determine a straight line. Accordingly in § 120 Mr. Russell tries back, and thinks he can now see the reason for what may previously have seemed 'a somewhat arbitrary fact,' namely the necessity of *four* collinear points for an anharmonic ratio. I quote the important part of his explanation—"Given one point. . . . . no projective relation to any second point can be assigned which shall in any way limit our choice of the second point. Given two points, however, there is such a relation, the third point may be given collinear with the first two. This limits its position to one straight line, but since two points determine nothing but a straight line the third point cannot be further limited." Is this really the case ? Why not limit it by saying it shall be between or outside the two given points ? But I am anticipating. Mr. Russell goes on : " Thus we see why no intrinsic projective relation can be found between three points which shall enable us from two uniquely to determine a third. With three given collinear points however we have more given than a straight line." . . . . Where? Do three given collinear points determine a plane? If not, how are you going to perform the quadrilateral construction without assuming something not given? And if you may assume a plane not given, why might you not do so before, and perform a trilateral or some other construction to determine the third point from the first two? " . . . . . and the quadrilateral construction enables us uniquely to determine any number of fresh collinear points." Precisely; anharmonic ratio depends on that beautiful quadrilateral construction, and the quadrilateral construction on anharmonic ratio, in a charming, if somewhat vicious, circle.

The point that Mr. Russell has missed is that if we are given a number of collinear points we *are* given something more than the relation that they are in a straight line. He says (§ 121), "Let us begin with a collection of points on a straight line. So long as these are considered without reference to any other points or figures they are all qualitatively similar. They can be distinguished by immediate intuition, but when we endeavour, without quantity, to distinguish them conceptually, we find the task impossible, since the only qualitative relation of any two of them, the straight line, is the same for any other two." The straight line however is *not* the only qualitative relation between the points taken together. In fact, as I shall show, the *straightness* of the line in 'projective geometry' is not a relation between the points in the line at all, but a relation between them and points outside the line. But there is another relation, or rather a system of relations, between the points themselves which permits us to say they are in a *line* and not spread over a surface. This system of relations is constituted by the *order* of the points in the line. If we name only two points, or as I shall show, even if we name three, we can not distinguish any *order* among them, but if we name four or more we can distinguish an order. And this order is what Mr. Russell calls a 'qualitative,' as distinguished from a quantitative, or even a numerical, relation. It *does* enable us to perform what Mr. Russell, in the sentence quoted above, describes as "an impossible task." It is not necessary to take into consideration any points external to the line to distinguish this order; and, unless we do consider such points, there is nothing to tell us whether the line is straight or not—the relation of straightness is a relation to points outside the line, not between the points in it. If Mr. Russell, or geometers generally, had attempted to define a 'line' qualitatively, that is to distinguish not a straight from a curved line, but a line of any kind from a surface, they would probably have been led to a completely different conception of the meaning of projective geometry. It is, in its abstract form, the science of the ordering, or cataloguing, of continuous groups, and the question whether 'projective' methods can or can not distinguish one figure from another is the question whether we can, or can not, distinguish the *order* of the points in one from the *order* of the points in the other; using this term in an extended sense which is applicable not only to groups of more than one 'dimension,' but also to continuous, as well as to discrete, groups.

PART II

THE ELEMENTS OF A SCIENCE OF CATALOGUING CONTINUOUS GROUPS.

1. I have shown elsewhere* that the ideal method of founding a science is to define the terms by verbal definitions by connotation, the primitive terms employed in the definitions being only such as are 'well understood'; and to proceed, in the first place, to a purely symbolic argument. We may always afterwards ascribe to them any real import which the terms of the science will bear, when we have sufficiently investigated the formal consequences of our definitions. If our primitive terms are not 'well understood' we shall sooner or later be led to contradictions or 'antinomies'; if that should happen there would be nothing for it but to try back, and make a fresh start. If, however, we come to no contradictions or 'antinomies' we may feel confident that our primitive terms really were as well understood as we took them to be. This confidence in our primitive terms may be expressed as a (subjective) axiom as soon as we have defined any terms which may be necessary to express the axiom clearly.

2. By *passing in review* logical units of thought, I mean thinking of them one after the other, or successively. If a collection of units has been *ordered* or arranged in order, I mean that it has been agreed that we shall pass the units in review in certain ways only, and not in others; so that after any one unit we shall, in passing in review, (or I may say simply in *passing*) think next of one of a certain number of determined units which may be spoken of as 'contiguous' to it, and not of any other unit whatever in the collection. Such an ordered collection I shall speak of as a *group*. I may now express the axiom of the science thus:

Any number of units may be conceived, discriminated between, and arranged in order.

* See my *Essay on Reasoning*.

By *collating* two groups, I mean regarding the order in each as identical, by taking one unit in one group to correspond to each unit in the other, and passing the corresponding units in review together.

When I say that two units are in different parts of a group with respect to a certain *boundary*, I mean that the units have been arranged in order, so that in passing in review from the one unit to the other in a different part of the group with respect to the boundary, we must pass through some unit or other which is *in* the boundary. Thus a boundary is itself a group of units.

I shall call a group *simple* if it does not itself form a boundary within itself. If we were to collate a group which was not simple with one which was, we should have to collate one or more of its units with two different units in the simple group.

I call a group *uniform* if the order proceeds in the same way from every unit in it. If it contains more than two units it must be possible to pass from one to a second through a third, and therefore it must be possible to pass through the third from the first to the second, and so on. Hence in any uniform group a single unit can not form a boundary.

An uniform group of the *first order* is one in which any two different units form a boundary; and a simple uniform group of the *order* $(1+r)$, where r is any positive integer, is one in which any simple uniform group of the order r forms a boundary, but one of a lower order does not.*

Thus sub-groups, as any groups of lower order than the whole group we are considering may be termed, are boundaries in the group of the order next above. As in a group of the first order two units form a boundary, we may call them a group of the order zero.

3. These definitions might be applied to discrete groups, that is to groups of a finite number of distinct units. The

* NOTE.—I should like to call attention to the fact that so far as I am aware, in no elementary text book (unless my " Foundations of Geometry " comes under that heading) is any proposition analogous to these definitions deduced or deducible from the premises. Euclid simply begs the question in his 11th Book.

theory of such groups, however, requires no special elucidation. What we are here concerned with are continuous groups, which may be thus defined—

When I call a group *continuous*, I mean that between any two units in it any finite number of boundaries may be conceived. To catalogue such groups it is necessary to have in them certain boundary groups. These boundaries must be defined in some way. We are not concerned to enquire *how* we recognise individual units, or *how* we know whether we can, or can not, pass from one to another without passing through a boundary. All we have to do is to establish a system of cataloguing which can be applied after these preliminary processes of recognition have been performed. And similarly it is not necessary to say how a boundary is to be recognised, if we lay down as a definition the conditions which such recognition shall imply. The system defined below is not a necessary or the only possible system, but it is *a* possible one, and that is sufficient for the purpose of the science.

The definition of the group of order zero might be accepted as a mere verbal definition. But I may as well point out the psychological reason for it. It is that in the only kind of complete group which we conceive as a whole, that is without intentionally or through ignorance confusing it with an incomplete group, two units are required to form a boundary. The geometrical conception of a closed line is obviously a case in point, but the same thing may be exhibited in non-spatial groups. If I regard the numbers 1, 2, 3, 4, 5 as forming an uniform group, I regard 5 as contiguous with 1 ; and consequently any single number, say 3, does not by itself form a boundary between two others, say 2 and 4. The subsequent developments of the theory of order among continuous groups may, I think, be held to show that this psychological fact is of universal significance, but for the time being we may regard the definition in question as purely arbitrary.

4. As I shall define 'unique' boundaries in terms of 'unique' boundaries of lower orders, the unique boundary of the order zero, which I shall write U_0, is defined as two units, either of which uniquely determines the other, and which may be called

'antipodal' units. The unique boundaries of higher orders may then be defined thus—

An *unique boundary of order* r, U_r (where r is an integer higher than zero and lower than n, the order of the whole group we are cataloguing) is uniquely determined by naming any unique boundary of the order $(r-1)$ (U_{r-1}), and one of the order zero, (U_0) not in the former, in the U_r.

5. It must be carefully observed, that though I define an unique boundary in terms of one of a lower order, there is nothing which I have laid down as yet to define the unique boundary of the lowest order at all. As far as this science is concerned the antipodal relation between units in a U_0, and the relations which determine the uniqueness of a U_1, are arbitrary. They must be determined by methods which do not come into the discussion of the science of continuous groups as such at all. And consequently there is nothing in the science to determine the whole collection as an 'unique' group of the order n; we can only say it is *a* group of the order n. In any particular application of the science our method of determining unique boundaries may be such as to define uniquely the whole group, but this depends on considerations foreign to the abstract science with which I am now dealing. This alone is sufficient to distinguish my theory from that of Veronese* for he takes the whole group to be defined by the sub-groups which it contains; and this shows, what indeed is otherwise evident, that Veronese's theory is not free from geometrical pre-conceptions; and it is not therefore surprising to find that its scope is more restricted than that of the theory I am now advancing.

And in the same way the choice of the 'order' of the whole group, or that of the highest unique boundary in it, is also determined by considerations foreign to the abstract science of cataloguing. I can illustrate this point at once by an analogy from the case of discrete groups, though of course the analogy can not be carried beyond a certain point. If we wish to be able to find any named book in a library, what we do is to prepare a *catalogue*. Now a catalogue is simply a group of

---

* See *Grundzüge der Geometrie*, the German translation of Veronese's work, p. 219 (Teubner, Leipzig). And also p. 70 of this paper.

units whose order is intuitively known. The simplest example is the group of natural numbers, and we might catalogue the library by simply numbering the books from 1 to as high a number as was necessary. It is not at all necessary that the numbers should be consecutive. If the book I want is number 3 and I see numbers 2 and 7 on the shelves, I know my book must be between them—if there is only one book between them either that is the one I want, or the one I want is not in the library. Whether there are or are not books corresponding to numbers 4, 5, 6 does not matter at all. In this way then the library may be catalogued as a (discrete) group of the first order. But I may also number the bookcases, the shelves in each bookcase, and the volumes on each shelf separately. In that case I have catalogued the library as a (discrete) group of the third order. I might catalogue the rooms separately also, and so make it a group of the fourth order. I might use letters of the alphabet instead of numbers, in some or all cases, for we also know intuitively the order of letters of the alphabet. The arithmetical properties of the numbers have nothing to do with it—it is their order only which is made use of.

6. I must add a few words of explanation as to what I mean by the unique determination of one group by others. This sort of phraseology is indeed common enough in books on projective geometry and elsewhere, but it needs more precise definition before it can be considered ‘well understood.’ If I say that a group C is uniquely determined by groups A and B, I mean not only that any group so determined by A and B *is* the group C and no other, but further that the fact that it was so determined by A and B *involves all that can possibly be said about it*; and therefore any conclusions about units in C, which flow from the fact that they are *in C*, must be deducible from the facts which determine A and B. And further, anything that can be said about units in C without referring to the determinations of A, must be deducible from the determinations of B, and *vice versa*. This is a general principle which applies not only to the particular cases of unique determination which I have already mentioned, in the definition of unique groups, but to other cases where units or unique groups are defined as

being common to two or more unique groups, or in other ways which come under the general head of unique determination. The principle will be amply illustrated later on.

7. If A, B, be two simple, but not necessarily unique, groups of the order $(r-1)$ which are boundaries in a group of the order r, and if two units in A are respectively in the two different parts into which the boundary B divides the group, then we can not pass from one of the units in A to the other, while passing in review the units of the group A alone, without passing through an unit in the group B. Therefore the units common to the groups A and B form a boundary in A, that is they form a group (or groups) of the order $(r-2)$.

8. For the sake of abbreviation I use the symbol U_r to denote an unique boundary, or group, of the order r, and I shall call the whole collection a complete group of the order n, or G_n though we may use the symbol U_n if it is clearly understood that in this case it is not defined, or determined, by unique groups of lower orders. The unique group of the first order is therefore a closed or complete group, and the lowest unique group is that of the order zero, and consists of two units, each of which determines the other uniquely, and which may be called *antipodal* units. Since an unit determines a U_0 uniquely, we may say a U_r is determined by a U_{r-1} and any unit not in it, instead of saying a U_0 not in it.

Any two different (U_0)s, or two different units which are not antipodal, (*i.e.* are not in a U_0) determine a U_1. Therefore any three (U_0)s, or any three units, not in a U_1, determine a U_1 and a U_0 not in it; and therefore a U_2. Similarly any four which are not in a U_2 determine a U_3. And generally, any $(r+1)$ units, or (U_0)s, which are not in a U_{r-1}, determine a U_r. and any $(p+1)$ units, or U_0s, which are not in a U_{p-1}, if they are in a U_r, determine a U_p which is wholly in that U_r. Thus if any two unique groups have a common unit, they have common an unique group of units. In particular, if any unit is in an unique group, so is its antipodal.

9. If a U_p and a U_q, A and B, have common a U_s, C, and no other units, then they determine a U_{p+q-s} (which contains both of them), uniquely. For if we take $(s+1)$ (U_0)s in C,

which are not in a U_{s-1}, and $(p-s)$ more (U_0)s in A, but not in C, and therefore not in B, and which, with the $(s+1)$ (U_0)s chosen in C, are not in one U_{p-1}; then in all we have $(p+1)$ (U_0)s in A, which determine A. If we now chose a (U_0) in B which is not in C, it determines with A a U_{p+1}; and with C a U_{s+1}, which is common to B and the U_{p+1}. If we chose another U_0 in B not in this U_{s+1}, it will determine with the U_{p+1}, containing A a U_{p+2}, and with the U_{s+1} common to the U_{p+1} and B, a U_{s+2}, which is common to the U_{p+2} and B. And so we go on, till we have chosen $(q-s)$ (U_0)s in B, and have a U_{p+q-s} which contains A, and a U_{s+q-s}, that is a U_q, common to the U_{p+q-s} and B, which is necessarily B itself.

Conversely if a U_p and a U_q are both in one U_r, they will have at least a U_{p+q-r} common, if $(p+q)$ is greater than r. For if a group of order s contained all the (U_0)s they had common, the U_p and the U_q would determine a U_{p+q-s}. But they cannot determine a group higher than r, since they are contained in a group of order r. Hence r must be at least as great as $(p+q-s)$, or s must be at least as great as $(p+q-r)$. If $p+q=r$ they must have at least a U_0 common; if it is less than r however, they may have no common units at all.

We may say that a U_p and a U_q *determine* an unique group common to them, in the same sense as that they determine an unique group which contains them. In the one case the group determined consists of the whole of the common units, and is not any group of lower order contained in it, and in the other case the group determined is the lowest order of group which contains both of them, and not any higher group.

10. Perhaps a modern philosopher would be inclined to see an 'antinomy' in the fact that though all we can say about an unit is that it is or is not in the same part of the group as another unit, with respect to a given boundary, yet I speak of units as being determined by being 'in' certain boundaries. The apparent contradiction is however easily explained away if we remember that a boundary is itself nothing but a group of units, which may be reciprocally defined by the units which are in different parts of the whole group, with respect to the boundary, just as well as the units may be by the boundary.

This is most readily seen in the case of a complete group of the first order. If a and c are in different parts of such a group with respect to b and d as a boundary, we may just as well define the boundary (bd) by means of a and c, as the unit a by means of the boundary (bd) and the unit c. Which is the one to be defined, and which the one already understood, depends on considerations outside the scope of a symbolic science; it must be determined for any particular application of the science on grounds with which formal logic has nothing to do. Thus we may always use units to catalogue boundaries, as well as boundaries to catalogue units; and the whole process of cataloguing consists as a matter of fact of a combination of the two processes. We may say of a that it is 'in' the boundary $abcd$, and not in the same part of it as c with respect to (bd); but this is only a first step in the cataloguing of a. We must go on to say $(abcd)$ is a boundary which is 'between' two units e and f, in a group of the second order, just as we might have said b was in a boundary of the order zero 'between' a and c.

The assertion that a boundary is 'between' two units is in fact merely the logical converse of the assertion that two units are separated by a boundary. The boundary has in this case become a logical unit of thought, though we reserve to ourselves the right of subsequently analysing it as a group of subordinate boundaries or of ultimate units.

11. If therefore we wish to catalogue a group of the order n, we may say of any unit in it (1) it is in an unique boundary of the order $(n-1)$ which is 'between' two known units; (2) in this boundary group it is in an unique boundary of the order $(n-2)$ which is between two known units...and so on; till we come to (n) it is in an unique boundary of the lowest order, zero, which is between two known units in a U_1; and lastly, it is, or is not, a known one of the two units in this group.

Now since in any continuous group we may take any number of boundaries between any two given units, the determination of a boundary in this manner can never be more than approximate. Though theoretically sufficient, in practice any system of cataloguing of continuous groups can never be more than practically sufficient for the purpose in hand, whatever that may be. But the degree of approximation reached will not depend on the order of the group; that only determines the number of approximate determinations. The degree of accuracy can be made to depend on the accuracy with which we can catalogue the units of a group of the first order, as I shall proceed to prove. At present I wish to call attention to the fact that to catalogue a group of the nth order we have first to catalogue n groups of unique boundaries, considered as logical units of thought for the time being. And, while doing so, any proposition which is proved generally as applicable to the cataloguing of units, applies *ipso facto* to the cataloguing of unique boundaries.

12. Without two distinct units we can have no boundary. This is obviously true in the case of the only kind of group we can clearly conceive as a whole, which is '*totus teres atque rotundus*'; of which a closed curve is an example. But perhaps it is not quite so obvious that it is equally true in any method whatever of cataloguing a group. If so, however, it is only because we confuse our minds by introducing the term 'infinity' into the discussion. Logically an unit or point 'at infinity' is not different from any other unit or point; it is only because we look at it from a different point of view that it appears to us different. If, for extraneous reasons, we have adopted this point of view, we must not let ourselves be led into making statements which would obviously be erroneous if we changed our point of view. That is why in this discussion I have adopted the point of view from which the group can be conceived '*totus teres atque rotundus.*' If we "project points to infinity" we are apt to lose sight of them, but if it is logically done it can not alter the logical relations between them. Consequently it is always true that it takes at least two distinct units of thought to form a boundary. And there is no sense in talking of a boundary unless it separates two units. Therefore it requires at least four units in order that we may recognise any *order* among them. This proposition is very important, and it applies to all systems of cataloguing alike.

13. If in any *complete* group of the first order we are given

two units which are not antipodal, we are at the same time
given two more, namely their antipodals. And as the antipodals
are uniquely determined by the units, we are given the order
of the four units. It may be shown that an unit and its
antipodal are in different parts of the group with respect to
any other unit and its antipodal, considered as a boundary.
Let aa' be two antipodal units in a complete group of the first
order (which may or may not be 'unique,' if it is a sub-group
in a higher group). Let us pass in review the units in one of
the parts into which (aa') divides the group, calling the unit
under review b, and its antipodal b'. If $a_1a_2a_3\ldots$ be a succession
of the units passed under review, and $a_1'a_2'a_3'\ldots$ their antipodals,
then as we successively collate b with $a_1a_2a_3\ldots$ we collate b' with
$a_1'a_2'a_3'\ldots$; and until b passes the boundary (aa'), b' can not do so.
Therefore b, b', are always in the same, or always in different,
parts of the group with respect to the boundary (aa'). But if
b' were in the same part as b, then $(b'a)$ would form a boundary,
with respect to which b and a' would initially be in different
parts of the group. And b reaches a' without passing through
a; it would therefore have to pass through b'; which, however,
it can not do, since an unit and its antipodal are by definition
distinct. Hence b, b', are always in *different* parts of the group
with respect to (aa'). Further, we see that if b passes any
unique boundary (a_1a_1'), b' also passes the same unique boundary.
Hence the order of units is the same as the order of their
antipodals.

14. If besides (aa') (bb') we are given a fifth unit, c, we can
say of it that it is in one of the four parts into which the two
(U_0)s divide the whole group. We may for example say that it
is not in the same part of the group with respect to the
boundary (ab) as either of the antipodals of a and b, or we
may make a similar statement with respect to (ba'), $(a'b')$ or
$(b'a)$.

In the same way in a complete group of the second order,
if we have a complete group of units which forms a boundary
in the group of the second order, and is such that the antipodals
of the units are all in one part of the group with respect to
this boundary, then the group of the first order, with its

antipodal group, divides the group of the second order into three parts, and we can say of any unit not in either group, either that it is not in the same part with respect to one of them as the antipodal group, or that it is in the same part, or that it is not in the same part, with respect to the antipodal group as the original group was. And similarly with groups of orders higher than the second.

15. Now I have said that in any system of cataloguing we can say of two units that they are separated by a boundary, which must itself consist of at least two units; and reciprocally we may speak of a group of at least two units which forms a boundary as 'between' two units. But we can not, without further definition, speak of a single unit as 'between' two others. In a complete group, however, a single unit uniquely determines an unique boundary, and we might speak of a single unit as between two others, meaning that the U_0 determined by it was between them.

I intend, however, further to limit the use of this expression. If I say unit a is between units b and c in a U_1, I mean not only that the (U_0) (aa') is between b and c, but that neither is the (U_0) (bb') between a and c, nor the U_0 (cc') between a and b. This is the same thing as saying that a is in a different part of the U_1, with respect to (bc) as boundary, from the antipodals of b and c. Thus, *in this special sense*, we can, *in a complete group*, speak of one unit as 'between' two others, but it must clearly be understood that this is only possible by an implicit reference to their antipodals, and that if there were no uniquely related pairs of units the expression would be meaningless. In the case of a bounded group we might indeed give another meaning to it by saying an unit was between two others if it was in a different part of the group from the boundary of the whole group, with respect to the boundary formed by the named units within the group. But in that case, though not referring to antipodal units, we should be referring implicitly to a boundary assumed arbitrarily in the group, the units in which would be specially related to all the other units in the group.

In the same way if a group of units with the group of antipodal units form two boundaries which divide a group of order higher than the first into three parts, we may speak of units as 'between' the two groups or 'within' either of them.

If, however, we take $(U_0)s$ as our units of thought, and consider a (U_1) as a group of the first order of such logical units, we shall require two of them to form a boundary; a single U_0 is not a boundary for $(U_0)s$. And conversely we can not say that one (U_0) is 'between' two others in a U_1. Nor, in a U_2, will a U_1 be a boundary for $(U_0)s$; and we can not therefore say that a (U_1) is between two $(U_0)s$; though it may be between two units, one selected from each. But two $(U_1)s$ form a boundary for a U_0 in a U_2; and therefore we may say that two $(U_1)s$ are between two $(U_0)s$, or that two $(U_0)s$ are between two $(U_1)s$. And the group of $(U_0)s$ which forms the two antipodal groups of units which divide a U_2 into three distinct groups of units, forms a boundary for $(U_0)s$, and divides the U_2, considered as a group of $(U_0)s$, into two distinct parts. We may therefore say of a U_0 that it is within or between the groups of units, according as the units in it are so, and though it sounds unfamiliar, we may retain the same expression and say a U_0 is within or between the single group of $(U_0)s$ formed by the two groups of units. Such a group of $(U_0)s$, which is a boundary in a U_2 for $(U_0)s$, I shall call a B_1. A group of units which contains the antipodal of every unit in it, of which a U_1 is a special case, I shall call an A group. Similarly among higher orders of groups we may distinguish B from A groups, according as they are, or are not, boundaries for $(U_0)s$. It will be observed that a U_r is always an A_r, though all $(A_r)s$ are not $(U_r)s$.

16. If we pass in review the units in any complete group, and in so doing pass any boundary, we must repass that boundary before we can return to the unit from which we set out. Therefore on the whole we must pass that boundary an even number of times. For example, either an A_1 or a B_1 in a U_2 must have an even number of U_0s common with a B_1 in it. And any complete group of units in it has an even number of units common with any other complete group. But to U_0s an A_1 is not a boundary in a U_2. Hence a complete group of U_0s may have an odd number of U_0s common with an A_1. If the

group is itself an A_1 it will always have an odd number; for if we trace an unit in one A_1, the other A_1 is a boundary for units and the antipodal of the unit under review must be in the opposite part of the U_2 with respect to this boundary. Therefore as we trace the unit under review from a given unit to its antipodal, it must pass the boundary an odd number of times. We shall at the same time have traced the U_0 from the initial U_0 back to it again, and the U_0 will therefore pass the boundary an odd number of times, and we shall have passed in review the whole group of $(U_0)s$. Similarly an A_1 has an odd number of $U_0 s$ common with any A group, an even number with any B group if they are boundaries for units in a group of higher order.

17. A group of $(U_p)s$ determined by a (U_{p-1}), called the *origin* of projection, and the various $(U_0)s$ in a U_q, which has no U_0 in the origin, is called a *pencil* of $(U_p)s$ (each of which is called a *ray* of the pencil) of the order q. The U_q is called the *initial section* of the pencil, and any other U_q in the U_{p+q} containing the whole pencil, which has no U_0 in the origin, (and which therefore has one U_0 common with each ray) is called a *section* of the pencil.

There are only two sorts of pencils which I intend to discuss—pencils of $(U_1)s$ of various orders, and pencils of various unique boundaries of the first order. By means of one or other of these we can reduce the question of cataloguing of units in any group to that of cataloguing of units, or $(U_0)s$, in a U_1. This may be done as follows:—

Take any two $(U_0)s$, O_1, O_2, and the U_1 determined by them. We may describe this as taking O_1 as origin, and O_2 as section, of a pencil of $(U_1)s$ of the order zero. Next take another U_0, O_3, not in this pencil; and take a pencil of $(U_1)s$ with origin O_1 and section (O_2O_3), that is the U_2 $(O_1O_2O_3)$. Then choose a fourth U_0, O_4, outside the pencil, and a pencil of the first order with (O_1O_2) as origin, and (O_3O_4) as section. Choose O_5, outside this pencil, and a pencil of the first order with $(O_1O_2O_3)$ as origin, and (O_4O_5) as section; and so on. In the end we have chosen $(n+1)$, $(U_0)s$, and $(n-1)$, $(U_1)s$ as sections of various pencils of the first order. But we may equally well follow the process in this way. Take an origin O_1, and a U_1 (O_1O_2) through it.

Through O_2 take any U_1 (O_2O_3) as section of a pencil of the first order of $(U_1)s$, with O_1 as origin. Through (O_2O_3) take any U_2 $(O_2O_3O_4)$ as section of a pencil of the second order of $(U_1)s$; and so on, till in the end we have a U_{n-2} determined by the n $(U_0)s$ $(O_2O_3...O_{n+1})$, and a pencil of the order $(n-2)$ of $(U_1)s$, with O_1 as origin. We want names for these two ways of collating, and an obvious analogy suggests that we should call them the Cartesian and Polar methods respectively. But the methods are not logically distinct. In the polar method we have to commence by collating together the rays of a pencil of $(U_1)s$ of the order $(n-2)$. In the cartesian method we have however to collate together the sections of the various pencils of $(U_1)s$ of the first order, and they form a pencil of $(U_1)s$ of the order $(n-2)$. In the polar way of looking at it we must collate together the rays of each pencil of $(U_1)s$, though these need not necessirily be collated with the rays of the next pencil. In the cartesian way of looking at it we must collate together the sections of the various pencils of any given order, but the sections of pencils of different orders need not be collated together; they may, if we like, be catalogued independently on different principles; just as in a library we may catalogue the book-cases by roman figures, the shelves in each book-case by letters of the alphabet, and the books in each shelf by arabic numerals. The next question to be investigated therefore is, how to collate together the rays of a pencil of $(U_1)s$ of any order.

18. I have now to prove a most important proposition, namely, that if we pass in review a continuous series of sections of any pencil of $(U_1)s$, each ray of which therefore has a single U_0 common with each section, and if we trace the unit common with the sections in each ray, from the unit in the initial to the unit in the final section, then the orders of the units so traced will be the same in the initial and final sections. I shall call this process *collation by projection*. It is sufficient for my present purpose to consider only the case of pencils of the first order, where the section therefore is a U_1; for as we have seen, the cataloguing of the whole collection may be made to depend on the collation and cataloguing of $(U_1)s$. The proof

however is general in its nature, and obviously extends to projection by pencils of U_is of all orders.

I have to show that in passing in review sections of a pencil of the first order of (U_1)s, either no unit in the section under review passes any boundary formed in the section by other units, or else that *every* unit passes *every* boundary so formed, in which case the order, though we may regard it as reversed, remains the same as before, in so far as it is determined by boundaries collated with each other.   For the principle of boundaries does not allow us to distinguish between 'passing in review' forwards or backwards.

Let oo' denote the origin of projection and $abcd \ldots a'b'c'd' \ldots$ the units in the section under review; $oao'a'$, $obo'b'$, and so on, denoting always the same rays, respectively.   Let suffixes to the letters denote corresponding units in particular sections, the suffix zero those in the initial section, the suffix n those in the final one.

Now bb' are always units in the ray $ob_0o'b_0'$, and this is a boundary to units in the U_2.   Hence any other unit in the section under review, say a, can not pass the boundary (bb') in the section, unless it passes through the boundary $ob_0o'b_0'$ in the U_2.   Now a is always in the ray $oa_0o'a_0'$, and the only units common to this ray and the ray $ob_0o'b_0'$ are the units o and o'.   Therefore a cannot pass the boundary (bb') in the section under review unless it passes through o or o', which it can only do if the section itself passes over oo', the origin of projection.   Consequently if the section does *not* pass over the origin, no unit in it can pass the boundary formed by any two other units, and therefore the *order* of the units collated by projection in the initial and final sections must be the same.

Next suppose the section *does* pass over the origin.   The initial and final sections must have a U_0 common, for they are (U_1)s in a U_2.   Let the one denoted by (a_0a_0') in the initial section be this common U_0.   And first let us suppose that, in passing from the initial to the final sections, the section under review always has the $U_0 (a_0a_0')$ in it, and further let b be an unit which passes through o as we trace it from b_0 to b_n.

Now those units in the ray $a_0oa_0'o'$ which are in the same part of it with respect to the boundary $(a_0 a_0')$, as o is, and those in the initial section $a_0 b_0 a_0' b_0'$ which are in the same part of it with respect to the same boundary $(a_0 a_0')$, as b is, form a closed group of units which divides the whole U_2 into two distinct parts, and we may say that as b passes along a ray of the pencil from b_0 to o it is always *within* this group, and b' is always within the antipodal group. As a_0, a_0' are units in this boundary group, any unit in the section under review in the same part of it with respect to $(a_0 a_0')$, as b is, will also be within the group. Now if any one of the units in the section under review, besides $a_0 a_0'$, were to pass through the initial section, they would all do so together. And the units in the same part of the section under review, with respect to $(a_0 a_0')$, as b is, must all pass into or out of the boundary group together. Consequently, if one of them, b, passes out through o, they have all got to pass out, and they cannot do so otherwise than through o. That is to say, all of them, with the possible exception of a and a', pass through o, and their antipodals through o', as the section under review passes over the origin of projection. And as they do so, each of them will cross each of the boundaries in the U_2 formed by the rays corresponding to the others, and will therefore cross each of the boundaries in the section under review formed by any two other units. Thus the order of all the units in the section under review, with the possible exception of a and a', will remain the same.

Now *primâ facie* we are unable to say whether a and a' pass through the origin or not. For though we can certainly say they do not do so unless the section under review passes over the origin, yet, if it does pass over we have no means of tracing a or a', since the section and the ray $a_0 o \, a_0' o'$ are then identical. But for the same reason we cannot say positively that they do not do so. We might therefore, merely for the sake of uniformity, say that when the section passes over the origin, a and a' pass through it, and therefore a_n is identical with a_0' and a_n' with a_0. And it can be shown that in fact we *must* say so, for the sake not only of uniformity, but of consistency. For there is no reason why the section under review should always pass through $(a_0 \, a_0')$, and if it does not happen to pass through it at the moment it passes over the origin of projection, it will never coincide with the ray $(a_0 o \, a_0'o')$, and in that case without doubt a and a' will pass through o or o'. Hence we are bound to say, that if in projection the section under review passes over the origin, then *all* the units, as we trace them from the initial to the final section, pass through the origin; and consequently their order in the section under review remains unaltered. It is to be noted that if the section under review does not pass over the origin, $a_0 a_0'$ are each collated with themselves. If, however, the section does pass over the origin, each is collated with its antipodal. If, however, we are not concerned to distinguish antipodals, or if we are only discussing $(U_0)s$, then we may say that the unit of thought common to the initial and final sections is, in a single projection, always collated with itself.

19. In the course of this discussion I have successively 'collated' b_0 with b_1, b_2, b_3, ... up to b_n, and while passing the sections in review I called the unit under review always b. The symbols with suffixes always meant particular units, a symbol without suffix meant sometimes one sometimes another unit. I may say that I *identified* the units with the symbols with suffixes, but I only collated them, temporarily, with the symbols without suffixes. When the symbols are 'identified' with the units, to collate two symbols is to collate the two units, and to 'identify' two symbols is to identify the two units. For example, if I say that b_0 is collated with b_n, I mean that the units with which these are respectively identified are collated together. If I say that a_0 is identified with a_n, I mean that the units are identified. On the other hand, if the symbols are only 'collated' with the units, to identify two symbols only collates the units. For example, b was collated first with b_0 and then with b_n. The 'b' in each case was identified with the 'b' in the other, but the result was only to collate b_0 with b_n, not to identify them. It will be seen therefore that the use of symbols to denote units is simply an example of what I have called 'collation,' and that what I call 'identification' with symbols is really only a collation which it is agreed shall be permanently maintained during the discussion in hand.

If our symbols have among themselves a well-understood

order, to collate or identify them with units is to ascribe that order to the units themselves. This collation with a well-understood order is what I call cataloguing, as distinguished from simple collation. How this cataloguing can be effected in continuous groups it is the object of this paper to enquire, but among discrete groups the process is sufficiently familiar not to require elucidation. For the present purpose it is sufficient to take as an example of a discrete group, whose order is ‘well understood,’ the twenty-five letters of the alphabet. In order that the group may be a ‘complete’ group, I include in it the same letters repeated, but with accents to distinguish them, and regard any letter and the same letter accented as antipodals. If we regard the letters with suffixes in the above discussion as cataloguing the groups of the first order in this way, then we always collate a_n with a_0, but whether we identify it with a_0' or with a_0, will depend upon whether the section under review does or does not pass over the origin as we trace it from the initial to the final section.

This point being settled, no generality will be lost if in future I assume (i) that the section under review always passes through the $U_0 (a_0 a_0')$, and (ii) that any other unit in it, by which, in addition to $(a_0 a_0')$, it is completely defined, and by which it may be said to be catalogued, passes only through units in one ray which are ‘between’ the units in the initial and final sections which are to be collated together, and passes through each of them only once. This convention will determine whether the section passes through the origin or not, according as the origin is or is not ‘between’ the units in any one ray which are to be collated together. It is therefore always possible to collate together the units in two given sections which have a common U_0 by a single projection; the section under review passing through, or not passing through, the origin, whichever we please; and the units in the final section collated with the units in the initial section in the one case, will be the antipodals of those collated with them in the other. Hence we infer that the order of units is the same as that of their antipodals (which however we knew already).

20. Suppose we wish to collate together certain units in the initial, and the same number in the final section, selected arbitrarily. We may suppose the units in the initial section to be catalogued by the symbols $a_0 \, b_0$ ... and so on, and $a_n \, b_n$ ... will denote the units we collate with them respectively by projection. To use these symbols to denote the arbitrarily selected units would be to ascribe to them a certain order, which we must not do, as they are to be selected arbitrarily. I shall therefore use capital letters for the units of the final section, and shall conceive these letters not to have any intrinsic order, so that they do not *catalogue* the units which they denote. To catalogue them we have to collate them with the already catalogued units in the initial ray. Now we can say of c_0 that it is 'between' a_0 and d_0, and that it is not 'between' a_0 and b_0. Hence it must necessarily be impossible to collate a_0 with A, b_0 with B, and c_0 with C, if C is 'between' A and B; and similarly it must be impossible to collate a_0 with A, c_0 with C, and d_0 with D, if C is *not* between A and D. Hence it can not always be possible to collate together arbitrarily three units in a complete group with three. But I shall show that it is always possible to collate two with two by a single projection, if neither of them are in the U_0 common to the initial and final sections, and by more than one projection in any case, and further, that it is always possible to collate two out of three with two out of three and the third with the given third, or with its antipodal. Thus in any case it is always possible arbitrarily to collate three U_0s in a U_1 with three others in a U_1, by one or more projections.

Let $(a_0 \, a_0')$, as before, be the U_0 in the initial section which is also in the final one. Then by a single projection we can not collate a_0 with any unit A, unless this unit is also in the initial section; but if it is, we can effect the collation, whether A is identified with a_0' or a_0, by tracing the section under review either through, or not through, the origin; which origin may be taken anywhere in the U_2 containing the sections, except in the sections themselves.

Suppose further we wish to collate b_0 with B in the final section. We can do so by choosing any origin in the $U_1 \, (b_0 B)$. If neither of the units in the origin is 'between' b_0 and B we

shall at the same time collate a_0 with A. If one of them is
'between' b_0 and B we shall collate a_0' with A.

If further we wish to collate c_0 with C, the origin must also
be in the $U_1(c_0C)$, and it is therefore definitely determined. If
one of its units is between b_0 and B, and also between c_0 and C,
then we shall collate b_0 with B, c_0 with C, and so we shall if
neither unit is between either b_0 and B, or c_0 and C. But in the
first case we collate a_0' with A, and in the second a_0 with A. If
one unit of the origin is between b_0 and B, and neither unit
between c_0 and C, we collate b_0 with B, but c_0' with C. And
similarly we can discuss the remaining possibilities. It is clear
enough from these illustrations how it comes about that we can
not always project three units on to three, but can always project two out of the three on to two, and the third on to the
third or its antipodal, when one pair is in the U_0 common to
the initial and final sections.

If no pair is in the common U_0, or if the two sections have
no common U_0, we shall require two projections to effect the
collation. By one projection on to a section determined by,
say, a_0 and B, we can collate a_0 with a_1, which is identified with
it in the intermediate section, b_0 with b_1, which is identified with
B or B', and c_0 with c_1. Then by a second projection we can
collate a_1, b_1, c_1 with A or A', B or B', C or C' as the case may be.

If the initial and final sections have two (U_0)s common, that
is, if they are coincident, and if no pair of units to be collated
are identical or antipodal, it will require three projections. But
in the end we can always collate by projection two units out of
three with two, and the third with the third or its antipodal.
If we do not care to discriminate between antipodal units,
or if we are only concerned with U_0s, we may say that we can
always collate three with three.

21. To anyone who thoroughly understood the expression
'unique determination,' as I employ it in this science, this would
be a foregone conclusion. It would be sufficient to say, "No
order is distinguishable among three units of thought however
we may group them; and since projection is simply an unique
method of collation, that is of comparing orders, it must be
possible by projection to collate together any three (U_0)s, and

any three others, two and two.   But since, in a complete group, the selection of any three units carries with it *ipso facto* the selection of three antipodals, there is a certain order distinguishable among them, namely by the relations with their antipodals. Hence it cannot always be possible to project three units on to three, unless we agree to overlook the distinction between antipodals."   This, I say, is a logically sufficient proof; but from a philosophical point of view it is interesting to see how the definitions work themselves out.   In the next proposition I shall however adopt at once the simple logical method, and leave the detailed investigation to be worked out by others.   The logical principle may be enunciated thus.   If a number of units α, β, γ ... are respectively determined by units a, b, c ... by means of relations which determine them uniquely with reference to certain other arbitrary factors, then any statement which can be made about α, β, γ ... must be deducible from statements made about a, b, c ..., or about the arbitrary factors, or both. And further, any statement which can be made about α, β, γ ... independently of any determination of the arbitrary factors must be deducible from a similar statement made about a, b, c ... and *vice versa*.

22. Now suppose the $(U_0)s$ abc... in a U_1, A, are projected by a pencil of U_1s with any origin, and that any other section of this pencil is again similarly projected from some other origin, and so on p times successively.   And next let A be projected again by a succession of q different projections, by pencils of $(U_1)s$.   If the corresponding pairs of rays in the pth and qth pencils have any common $(U_0)s$ let them, or as many of them as there may be, be denoted by greek letters corresponding to the latin letters in the group A.   And let the group of $(U_0)s$ so determined be called the group X.

If the group X consists of a single U_0, α, corresponding to a in A, then there is nothing that can be said about a by itself, without reference to the origins of projection, nor is there anything that can be said about any other U_0 in A with reference to a, without reference to the origins of projection. Even if X consists of two $(U_0)s$ alone, α and β, there is nothing which can be said about them by themselves, nor is there anything that can be said about any other U_0 in A with reference to a and b alone, without reference to the origins of projection. There is therefore no reason why the group X should not consist of two (U_0)s alone.

But if there is any third U_0, γ, in X corresponding to c in A, then we can make a statement about c; namely that it is in a group of the first order in which (ab) forms a boundary. Hence γ must be in a corresponding group of the first order, namely, X, in which $(\alpha\beta)$ forms a boundary. And of any other (U_0) d in A we can say that it is, or is not, in the same part of A with respect to the boundary (ab) as c is. There must therefore be in X a corresponding U_0, δ, of which we can say that it is, or is not, as the case may be, in the same part of X with respect to $(\alpha\beta)$ as γ is. That is, if X contains as many as three (U)s it must be a group of the first order and contain a U_0 corresponding to each U_0 in A. We can not, however, say that X must be an *unique* group of the first order, for as I have pointed out there is no sense in calling A unique, except in reference to units in a wider group which contains it. Hence whether X is unique or not depends on units outside A, and may be affected by our choice of origins of projection. If, however, the pth and qth pencils are not in a U_2, they have a U_1 common if they have more than one U_0 common. Therefore if three pairs of rays have common (U_0)s, X can only be this common U_1; and as every pair of rays have a common U_0 in this U_1 it is a common section of the pth and qth pencils.

If, however, the pth and qth pencils are in one U_2, X may be a group of the first order, but not an unique group. In that case let ϵ be an *unit* (not U_0) in X corresponding to an *unit* e in A. Then in the U_2 containing X the $U_1 (\alpha\beta)$ forms a boundary for units, and we can therefore say of ϵ that it is, or is not, in the same part of the U_2 with respect to $(\alpha\beta)$ as another given unit is, or else that it is in the boundary $(\alpha\beta)$. We can correspondingly say that e is, or is not, in the same part of A, with respect to (ab) as a boundary, as a given unit is; but we can *not* say it is in the boundary (ab). Hence if one other unit ϵ in X is *not* in the $U_1 (\alpha\beta)$, there can not be *any other* unit in X which *is* in it, and conversely if ϵ is in $(\alpha\beta)$, *every* other unit in X must be in it also. That is to say, if three units, no two of which are antipodal, or if three (U_0)s in X are in a U_1, then X *is* that U_1, and is a common section of the pth and qth pencils, in which every pair of corresponding rays determines a U_0.

Now X must be either an A_1 or a B_1, that is, either it consists of a single group of units containing the antipodals also of each unit, or it consists of two distinct groups of units, one of which contains all the antipodals of units in the other. For if some and not all of the units had antipodals in one part of the group, we should be able to say something about these units which we could not say of the others, without reference to any units outside X, and we could not say anything correspondingly about the units in A. It is easily seen that if X is a B_1 as we pass in review the units in A, the corresponding units in X will all be in one part of the group till we come to an unit uniquely related in some way to the origins of projection. This will be the unit in X, which is in the U_1 determined by the pth and qth origins.

Since an A_1 must have an odd number of (U_0)s common with any other A_1, if X has only two U_0s common with any U_1 (which is a special case of an A_1), it must itself be a B_1 and not another A_1. This particular kind of B group, which is further specialised by the connotation that it can only have two (U_0)s common with any U_1, I shall call a C group of the first order, or C_1.

23. We may in the same way investigate the nature of the group X in the case of successive projections of the units of a U_2, A, by pencils of (U_1)s. If there are only one or two (U_0)s in X, then as before we can say nothing about them; nor can we do so if there are only three, unless the three (U_0)s in A, to which they correspond, are in one U_1, which would be the case just discussed. For if there are three (U_0)s α, β, γ, in X corresponding to a, b, c, in A, which are not in one U_1, we are unable to make any statement in respect to a fourth U_0, d, in A with reference to a, b, c; *for a U_1 is not a boundary for (U_0)s in a U_2*, and we can not therefore say that d is, or is not, in a different part of the U_2 from a, b, or c. But any two of the

$(U_1)s$ (ab), (bc), (ca) form a boundary in A, and therefore if there is a fourth U_0, δ, in X corresponding to d in A, we can say of any fifth U_0, e in A, that it is, or is not, in the same part of the U_2 as d is, with respect to one or other of these boundaries. Hence in this case it follows that X is a group of the second order, in which $(\alpha\beta)$, $(\beta\gamma)$, $(\gamma\alpha)$ determine boundaries. This group must be a U_2, and a common section of the pth and qth pencils. For α, β, γ can not all three be in one U_2 containing the pth and qth origins of projection, since a, b, c were not in one U_1; and if δ were not in the U_2 determined by α, β, γ, then, as before, we could show that no other U_0 in X, besides α, β, γ, could be in this U_2. But if X is a group of the second order, it must have at least a group of the first order common with the U_2 determined by (α, β, γ). Hence, if there are four $(U_0)s$ in X, which correspond to four in A, which are not in one U_1, then they are in a U_2; which is a common section of the pth and qth pencils. And similar results may be deduced in the case of successive projections by pencils of $(U_1)s$ of higher orders.

24. These results are very important. They might perhaps have been deduced by methods more familiar to students of projective geometry, though as a matter of fact I am not aware that any exactly corresponding propositions are to be found in the ordinary text books of the subject. It would probably be possible to deduce them indirectly from the proofs of the uniqueness of the quadrilateral construction, though the significance of these proofs appears to have been misunderstood by most writers on the subject. But from the philosophical point of view the proofs I have given are much more interesting, as they enable us to trace the true meaning of the results. The reason that we can not project any four collinear points on to any other four is to be traced, not to the fact that it takes two points to define a straight line, as Mr. Russell supposes, but to the fact that it takes two units, or $(U_0)s$, to form a boundary which shall separate two other units, or $(U_0)s$, and consequently we can make a definite statement about the order of four units, or $(U_0)s$, but not about that of any smaller number. It will, however, make this clearer to put the results in another way. We may say that there exists a

definite relation among any number of units, or $(U_0)s$, in a U_1 greater than three, which remains unaltered by projection. Similarly there exists a definite relation among any number of units or $(U_0)s$ in a U_2, greater than four, unless all or all but one of them are in a U_1, which is unaltered by projection. Now I have already shown that in its ordinary sense, with which we are familiar in dealing with discrete groups of units, the 'order' of units, or $(U_0)s$, is unaltered by projection; and further, that this is so even in the extended sense in which I have already used the word, that, namely, in which it is defined by the relations of units or $(U_0)s$ to boundaries, and which applies to groups of a higher order than the first. This new relation which we have discovered, which is unaltered by projection, is in fact nothing but the 'order' of the units or $(U_0)s$ in a new sense, or rather in an extension of the old sense which can be applied to continuous, as well as to discrete, groups. It is because no 'order' is distinguishable in a collection of less than four units, even if we regard it as a group of the first order, that we can always project three $(U_0)s$ in a U_1 on to any three other $(U_0)s$ in a U_1. And it is because an order is recognisable among three units, or more than three $(U_0)s$, that we can not always project three units in a 'complete' group on to three, or more than three $(U_0)s$ on to more than three; and that in either case, when we have projected three on to three, the remainder of the units in the initial group of the first order are uniquely projected on to, or collated with, the units in the final group. We shall become more familiar with this extended sense of the term 'order' as we proceed, but it is already clear that it is a sense which is capable of exact definition. It is no use 'rhapsodising' loosely about 'qualitative identity' or calling a relation which, *ex hypothesi* is non-metrical, 'anharmonic ratio.' If we can not strictly define our terms they serve at best only to disguise our ignorance, and it is more than probable that their employment will, sooner or later, lead us into positive fallacies, of the kind known to logicians as 'fallacies of the ambiguous middle.'

25. If in a complete group we collate the rays of a pencil of $(U_1)s$ by means of the units in each in the section, we shall

collate each ray twice over, with each of the two antipodal units. To evade this difficulty we must collate together the half-rays as divided by the origin of the pencil, collating together the units in the origin, and two other units in each half-ray. This would require the arbitrary selection of two units in each half-ray, besides the origin; but by the method I am about to describe it is sufficient to chose two units in the whole ray, one in each half of it, to collate a single ray, and only two more, besides the origin, in one other ray of a pencil of the first order, to collate the whole of the rays of the pencil; and, generally, $2n$ units besides the origin, to collate a pencil of the order $(n-2)$.

I shall call this process of collation of half-rays in a pencil *reversion*; and in reference to the process I shall speak of the origin as the *pivot* of reversion, and the section of the pencil as the *equator* of the reversion. It will appear that reversion is only a specially determined kind of projection, it involves no new principle.

26. The simplest case of reversion is where we are concerned only with two $(U_0)s$ in a U_1. We have seen that if the origin of projection is not in the U_1 and the final section coincides with the initial one, every unit is projected into its antipodal. But if we take the origin of projection *in* the initial section, the collation of all units in it, with the exception of the origin, and the units in the U_0 which is supposed to remain always in the section under review, is indeterminate. The unit in the origin may however be traced along any one of the rays of the pencil to its antipodal. At the same time, since the section does not pass through the origin, the units in the U_0 through which the section under review always passes, are collated with themselves, not each with its antipodal. Hence we collate a, b, a', b' with a, b', a', b respectively; that is, we can always regard the order a, b, a', b' as identical with the order a, b', a', b, *if* we regard aa', bb' as pairs of antipodal units. If these four units constituted the whole collection, and we catalogued the collection as a complete group by the letters a, b, a', b' this antipodal relation would *ipso facto* be established, and what we have proved only shows that it makes no difference to the 'order' of the letters

whether we read them forwards or backwards. If however we catalogued the group as $aa'bb'$, then (ab), $(a'b')$ would be the pairs of antipodal units, and in that case we could not collate a, b, a', b' respectively with a, b', a', b. The term '*singular reversion*' will do to denote this special case of projection.

27. I go on to the more general case of the reversion of the whole of the units in each of the two halves of a single ray. Let PP' be the pivot of the reversion and QQ' the equator. That is to say, P is to be collated with P, and P' with P', in each half ray, but Q in one is to be collated with Q' in the other, and *vice versa*. Take any U_1 through PP', and in it choose any two origins of projection $(O_1 \, C_1')$, $(O_2 \, O_2')$. To fix our ideas let O_1, O_2, be in the same part of the U_1 with respect to (PP'), and let O_2 be 'between' O_1 and P.

Let D be an unit in the U_1 between O_1 and O_2. If we project the units of (PQ) on to a U_1, (QDQ'), from one origin, and then project them back on to PQ from the other origin, P and P' will always be collated with themselves, and Q will always be collated with Q' and *vice versa*, for in one case the section will pass over the origin of projection, and in the other not. The other units in PQ will however in general be projected differently, according as we project first by O_1 then by O_2, or first by O_2 and then by O_1. If however *any one* of them is projected on to the same unit in either case, then, as we have seen, they will *all* be projected on to the same unit in either case. It is easily seen that any unit between P and Q will be projected on to an unit between P and Q', and that the 'order' of the units from P to P' *via* Q will be the same as the order of the projected units from P to P' *via* Q'. Thus we shall have uniquely 'reverted' the half ray PQP' on to the half ray $PQ'P'$. It only remains to be shown that an intermediate section QDQ' can always be determined which will effect this, after we have chosen the origins $O_1 \, O_2$; and that our choice of the origins O_1 and O_2, so long as they are in a U_1 with P, does not affect the result.

It will simplify the discussion to neglect for the moment the distinction between antipodal units, and regard unaccounted letters as denoting $(U_0)s$ instead of units. In this case it is evident that the distinction between pivot and equator disappears; either may be regarded as pivot and the other as equator—the reversion will be identical in either case. In order to make my treatment of the question more readily comparable with Mr. Russell's I will adopt the notation he employs in § 113 of his book, merely translating his paragraph, as literally as I can, into my own language.

' Given any three $(U_0)s$ A, B, D in one U_1, the quadrilateral construction finds the U_0 C into which A may be reverted with $B, D,$ as pivot and equator,' and (as I shall show) it also finds the U_0, C which, together with A as pivot and equator, reverts B into D, and further, shows us how to determine the intermediate section, by means of which all the $(U_0)s$ in a U_1 may be reverted with respect to a given pivot and equator. However, to continue the translation, ' Take any U_0, O, not in the U_1, ABD, and which determines $(U_1)s$ with B and D. Take any U_1 through A, having $(U_0)s$ P and Q common with the $(U_1)s$ (OD), (OB), respectively. Let R be the U_0 common to the $(U_1)s$ (DQ), (BP), and C the U_0 common to the $(U_1)s$ (OR) and (ABD). Then C is the U_0 required.

' To prove this, let T be the U_0 common to the $(U_1)s$ (DRQ) and (OA), and S the (U_0) common to the $(U_1)s$ (AR), (OD). Then by projection with origin R and section (OD), we collate A, B, C, D with S, P, O, D; and by a second projection, with origin A and section DQ, we collate them with R, Q, T, D. But again by a third projection with origin O and section (ABD) we collate them with C, B, A, D. Thus the order $ABCD$ is the same as the order $CBAD$.'

28. Now at first sight this conclusion seems simply *banal*. The two orders are evidently the same, considered as discrete orders of letters, except that one is written backwards. The point is, however, that it is not only the *discrete* order of the four U_0s which is the same, but the *continuous* order of the *whole of the* $(U_0)s$ *in* ABD, which have been projected at the same time. The simplest way to see this is to regard C as pivot, A as equator, and (AQP) as intermediate section, O, R being the origins of projection. In this case, projecting first by O as origin, if L be the U_0 common to (OC) and the intermediate section, we collate A, B, C, D with A, Q, L, P, and then with R as origin we

collate them with A, D, C, B. If we project by R first, we collate A, B, C, D with A, P, L, Q, and then with A, D, C, B, obtaining the same collation in each case.

If any reverted order be projected, it projects into a reverted order, taking corresponding units or $(U_0)s$ as pivot and equator. Hence it can easily be shown that C reverts into L with O, R, as pivot and equator. Consequently, to return to the former discussion, if we wish to revert a U_1, $PQP'Q'$ with respect to PP' as pivot and QQ' as equator, taking $O_1 O_2$ as origins of projection, to find D we perform the 'quadrilateral construction,' so that P reverts into D with $O_1 O_2$ as pivot and equator.

But now let us discriminate between the units in each U_0. If the letters in Mr. Russell's discussion represent only units, instead of $(U_0)s$, we can not say that the order $ABCD$ is the same as $CBAD$; for, wherever the antipodals may come in, they will serve to discriminate between the orders, *unless* (AC), (BD) are themselves pairs of antipodal units, which would be the case of singular reversion already discussed. It is, however, obvious that in one of the projections he makes (the first in his diagram) the section passes over the origin of projection, and in the other two it does not. Hence, what he has shown in the case of his diagram is that the order $ABCD$ is the same as $CBAD'$, if we take the letters to denote units in a complete group. Thus we discriminate between the pivot B and the equator D of the reversion. Looking at the reversion in the other way, as obtained by only two projections from R and from O, in either order, in one projection the section passes over the origin (namely when R is the origin, in Mr. Russell's diagram) and in the other not. Hence A reverts into A', and C is the pivot and A the equator of the reversion.

29. We may determine a reversion by naming the pivot, say C, and two units B and D, which shall revert into each other. We have in this case clearly named three units in the initial and three in the final sections, considering the reversion as a projection. But if we name the pivot and equator we have also named three units in each, namely C, A, and A' in one, C, A', and A in the other. For in reversion, not only do we collate B in the initial with D in the final section, but also D in

the initial with B in the final. Similarly, not only do we collate A in the initial with A' in the final, but A' in the initial with A in the final section. The case where the equator is named is a limiting case of the other, but it does not really by itself suffice to determine the reversion, for as we have seen in singular reversion the pivot and equator are named, but the collation of the other units remains indeterminate. The additional assumption implied in naming the pivot and equator of a definite reversion is however merely that the other units shall be reverted *somehow*; if they are reverted at all they are reverted in an unique manner.

We have seen that, so long as we are concerned only with $(U_0)s$, the pivot and equator need not, or rather can not, be distinguished. But even if we are dealing with units, the construction I gave may be used whether P is regarded as pivot or Q. P was made pivot by the fact that we chose that the section should cross the origin in one projection and not in the other. If it is made to cross it both times, or neither time, Q will be collated with Q, and P with P', so that Q will be the pivot and P the equator.

We have seen also that if we consider only four units A, B, C, D in Mr. Russell's construction, we may regard either A and B as pivot and equator, or C and D. The results as regards other units in the U_1 will however be totally different in the two cases. Reversion of a U_1 may be called a function of two U_0s, symmetrical with respect to them, or a function of two units, though not symmetrical with respect to them; but it is not a function of four, rather than of the infinite number contained in a whole U_1.

30. But if we are careful not to attach to it any 'metrical' or spatial significance, we may usefully employ the term 'symmetrical' in another way with respect to reversion. We may say that in the case of complete groups it defines an order symmetrical with respect to the pivot, and in which the units of the equator occupy symmetrical places. In the case of a group of $(U_0)s$ of the first order, the order defined is symmetrical with respect to either pivot or equator, but in a pencil of $(U_1)s$ the symmetry with respect to the equator, though it still exists,

is not of the same nature as that with respect to the pivot, for the equator is a U_1 or higher group, and the pivot a (U_0).

31. I return to the question of collation of boundaries. We have to arrange a system by which the units in the various half rays of each pencil of (U_1)s shall be collated in the simplest manner, that is, with the fewest arbitrary assumptions. It will be sufficient to show how the units in a pencil of the first order of half U_1s may be collated with each other, and with the section of the pencil. We are already given two U_0s in each ray, namely the origin and the U_0 in the equator. It will obviously be simplest to collate together the U_0 in each ray which constitutes at once the origin and the dividing boundary of each half ray. As the unit in the section of the pencil belonging to each half ray is used to collate the half rays, these units must all be collated together, and therefore *we must revert the units in each ray with respect to the origin as pivot and the section as equator* by an unique reversion, which will completely collate the units of the two halves of any one ray. If $(O_1 O_1')$, $(O_2 O_2')$, $(O_3 O_3')$ be the first three arbitrarily selected U_0s out of the $(n+1)$ which determine the system of pencils used to collate the whole group, $(O_1 O_1')$ will be the first origin and $(O_2 O_3 O_2' O_3')$ the first section; $(O_2 O_2')$ will be the second origin, and $(O_3 O_3')$ the U_0 with one of whose units the whole pencil of U_1s will be collated in the next stage of the process. Hence, to collate the two halves of the section with each other we similarly revert it uniquely with respect to $(O_2 O_2')$ as pivot and $(O_3 O_3')$ as equator. To collate together the section and the rays of the pencil, if we wish to do so, we may collate together the pivots of reversion and the equators in each case; thus O_1 in the ray is collated with O_2 in the section, and O_2 in the ray with O_3 in the section. Or we may catalogue the rays and the sections independently. To determine the collation completely, we have, however, to collate together some third unit in each ray and in each section. I will show how these third units may be determined for all the rays of a pencil of the first order by choosing arbitrarily units in two of them, or an unit in one and an unit in the section.

Select any two units, P_1 between O_1, O_2, and P_2 between O_1, O_3, in two rays of the pencil. Let R be the unit common to $(O_3 P_1)$ and $(O_2 P_2)$, and between the units named in each. Let Q_1 be the unit common to $(O_1 R)$, $(O_2 O_3)$ between O_2, O_3, and p_2 the unit common to $(P_1 Q_1)$, and $O_1 O_3$ between O_3 and O_1'. Next, revert the $U_1 (O_1 O_2)$ with respect to P_1 as pivot, so that O_1 reverts into O_2; and let any unit X between O_1 and P_1 revert into an unit Y between P_1 and O_2. And let Z be the unit common to $p_2 X$ and $P_2 Y$, between the units named in each case.

Now if we pass in review a series of units in the place of X, we get a series of units in the place of Z. And these units are common to corresponding rays of two pencils with origins P_2, p_2, and with sections which have themselves been projected from one and the same U_1. For Y was obtained from X by reversion, which is only a special case of projection. Therefore the locus of Z is either a U_1 or a C_1. It is easy to show that it is the latter, and that it passes through P_1, P_2, and p_2', which is the unit into which P_2 reverts with O_1 as pivot and O_3 as equator. For Z is within the boundary $O_1 O_2 P_2$, since it is between P_2 and Y, and Y is between O_2 and O_1. And if we trace X to O_1 at the same time we trace Y to O_2 and therefore Z to P_2. And if we trace X to P_1 we trace also Y and Z to P_1.

The locus of Z will have one unit common with every half ray in the pencil with origin O_1, and section $(O_2 O_3)$. If we collate together all these units, and the units in the origin and section of the pencil, we shall completely determine the collation of each half ray. And it can easily be shown that the loci of all units collated together are $(C_1)s$, no two of which will have a common U_0. We may imagine the collation to be effected by passing in review the rays of the pencil as sections of the group of $(C_1)s$, and tracing the units to be collated together along these $(C_1)s$. This is what I shall call *collation of the rays of the pencil by reversion.*

32. To determine the collation, besides the units O_1, O_2, O_3, we chose two more arbitrary units P_1, P_2. We might instead have chosen the units P_1, Q_1, arbitrarily; that is, one in the initial ray and one in the initial section. If in the next stage of the process we choose a fresh arbitrary unit P_3 in $(O_1 O_4)$, it will determine a fresh unit in every section. If on the other

hand we choose Q_1 and another arbitrary unit Q_2 in $O_3 O_4$, it will determine the arbitrary unit in every ray of the pencil with origin O_1. Thus in the end we have only to choose (n) arbitrary units $(P_1 P_2 \dots P_n)$, or $(P_1 Q_1 Q_2 \dots Q_{n-1})$ besides the $(n + 1)$ units $(O_1 O_2 \dots O_{n+1})$, to determine the collation completely. If we look at the whole collation from the polar point of view, we should naturally choose the units $(P_1 P_2 \dots P_n)$ arbitrarily. If on the other hand we look at it from the cartesian point of view, it would seem most natural to begin by selecting $(P_1 Q_1 \dots Q_{n-1})$. But it is of course logically indifferent which we do. If we choose $(P_1 Q_1 \dots Q_{n-1})$ arbitrarily in the initial sections, we are choosing the corresponding units in all the other sections by collation by reversion. In fact, the sections in the cartesian way of looking at it are really rays of pencils, and the rays in the polar way of looking at it their sections.

33. There is one property of unique boundaries which, though not very directly connected with the present discussion, deserves notice, not only on account of its intrinsic beauty, but because of the prominent place it occupies in most works on Projective Geometry ; and this is what is known as the Principle of Duality. We have already seen that we may say that a U_p and a U_q determine a U_s common to them in the same sense in which they determine a U_{p+q-s} containing them. And generally, we may say a U_0 is determined by $n (U_{n-1})s$ just as $n (U_0)s$ determine a U_{n-1}. To every proposition about $(U_0)s$ will correspond one about $(U_{n-1})s$, and so on. It must be noticed, however, that the principle of duality applies primarily to the whole considered as a group of $(U_0)s$, and not considered as a complete group of units. Now I have several times already spoken of considering the group as a group of $(U_0)s$ instead of one of units, and the time has come to consider more exactly what this means.

34. I will take first the simplest case, that of a group catalogued as a complete group of the first order of units, each unit of which determines another, which together with it forms a U_0. Now we have seen that in cataloguing such a group of units by means of unique boundaries we in the first place, as it were, run to earth a U_0, and then dig out of it the particular unit we want.

But suppose we stopped at the U_0, and so far from attempting to dig out a particular unit we regarded it logically as an unit in itself. In other words, suppose we started by cataloguing the complete group *without* establishing any unique relation between pairs of units, but retained all the rest of my investigation, in so far as it does not discriminate between the units in a U_1 as it stands. This would indeed logically have been the best way of starting the science, only it would have involved a re-investigation when we wished to get at what I have called the ' complete ' method of cataloguing, whereas by the plan I have adopted, the system in which there is no antipodal relation, which I will call the *Monistic* system, is really being investigated all the time alongside of the complete system. We have only to say that the $(U_0)s$ *are* the units of our collection, and our conclusions about them apply at once.

35. In the monistic system, if we pass in review the units, that is $(U_0)s$, in a U_1, we come back to the one with which we started, and so we do in exactly the same way if we pass in review the rays of a pencil of the first order of unique groups of any order whatever. In a U_1 it requires two units, *i.e.* $(U_0)s$, to form a boundary, and we cannot say that one U_0 is ' between ' two others. This introduces a complication into the method of cataloguing, for we can no longer identify an U_0 or any unique group, even approximately, by saying it is ' between ' two units; these being, on the M system, $(U_0)s$. We can only say it is, or is not, in the same part of the group with respect to these two $(U_0)s$ as a third named (U_0) is. It would be interesting to go on and investigate how this might most simply be done, but to do so would be beyond the scope of the present paper. I will merely call attention to a few more salient peculiarities of the M system. A single unit, (U_0), may be finally determined by two $(U_1)s$ in a U_2, or by n $(U_{n-1})s$ in the whole collection. An A_1 does not divide a U_2, nor an A_{n-1} the whole collection, into two parts, for it is not a boundary to the units of the M system. But a B group of the same order is a boundary, and divides the group into two parts which may be distinguished from each other by the fact that in one we can not have a whole A group of any order, whereas in the other we may. Any A_1 has an odd

4

number of units common with any A_{r-1}, but an even number common with any B_{r-1} in a U_r. Hence an A_1 must have at least one unit common with any A_{r-1} if both are in one U_r, but it need not have any at all common with a B_{r-1}. We can always collate by projection any three units, *i.e.* $(U_0)s$, in a U_1 with any other three. The principle of duality applies without any reservations.

36. In a complete group on the other hand, that is one catalogued on what I shall call the U system, if we pass in review the units of a U_1 we reach first the antipodal of the unit from which we started, and then, in order, the antipodals of all the units already passed in review, till we come back again to the initial unit. If, however, we pass in review the rays of a pencil of the first order, the case is different, for these rays, considered as units of thought, are catalogued monistically. To make the catalogue one on the U system we have to take half rays, divided by the origin of the pencil. In a U_1 two units, or one U_0, form a boundary. Hence we can say a U_0 is 'between' two units. We can also, in the special sense I have defined, say one unit is 'between' two units, if these are not themselves antipodal. Similarly, any unique group of a higher order, being a boundary, can be said to be 'between' two units. A B group is a double group of units, the two parts being antipodal to each other, and it divides the group in which it forms a boundary into three parts, and it may be said that an unit is 'within' one or other of the antipodal groups, or 'between' them. An A_1 has an odd number of $(U_0)s$ common with any A group, and an even number common with any B group, as in the M system. We cannot always collate three units in a U_1 with three others, but we can always collate two and two, and the third with the third or its antipodal. The principle of duality applies to unique boundary groups, that is a U_0 corresponds to a U_{n-1}, and so on, not an unit with a U_{n-1}.

37. We have now to consider certain other ways of cataloguing groups. There is one in particular which combines certain of the advantages of both the M and U systems, in that it is not necessary to establish an antipodal relation between

every pair of units, and on the other hand we are still able to speak of an unit or an unique group as 'between' two units. But to explain this I must go back to my original definitions and consider what is actually done when we establish the antipodal relation between two units. Suppose I have a collection of four units, and arrange them as a group of the first order, *abcd*, *d* and *a* being considered contiguous. Here *b* and *d* form a boundary which separates *a* and *c*. That is to say, if I pass in review from *a* to *c* I must pass through *b* or *d*. But why not pass straight from *a* to *c*? There is no unit to be passed through, no *boundary*. The answer is that I have agreed, by cataloguing the collection as a group of the first order, not to do so. I may say I have put a *barrier* between *a* and *c*, and also between *b* and *d*, in addition to the boundary formed by (*bd*) in the one case and (*ac*) in the other. Why, or how, I put the barrier it is not for formal reasoning to enquire. Now, there being an even number of units in my group, I may, if I please, just as I erected a barrier, forge a connecting link between *a* and *c*, *b* and *d*, and call them respectively antipodal units. If, however, my collection consisted of an odd number of units, I could not in this way establish an unique relation between pairs of units. I am at present unable to say how far this analogy may be carried in the case of continuous groups, or whether we ought not to say that though cataloguing on the *U* system undoubtedly implies an even number (however great) of units, yet cataloguing on the *M* system does not necessarily imply an odd number. But however this may be, the number of units in the *U* system being even we may divide it by two, and consider a collection of either an even or odd number of units to be catalogued as one half of a *U* group of any given order. I shall call this the semi-complete, or *S* system of cataloguing.

38. The simplest way of explaining it is to take the polar way of looking at a catalogued *U* group of the order *n*, and divide it in two by the section, or as I may call it the equator, of the pencil of $(U_1)s$ with origin (O, O_1'); and consider the *S* group to be formed of those units only which are in the equator or in the same part of the *U* group as O_1 is. It will then be seen that in the *S* group there is an antipodal relation only

between units in the equator, which will form an infinitesimally small portion of the whole continuous group, and which may for many purposes be neglected altogether. The conclusions we reached in the discussion of the U system may now be applied to the S system, by making certain provisos and reservations. In the first place we observe that whenever we have 'run to earth' a U_0 in the catalogue of the U group, if that U_0 is not in the equator, only one of its units is in the S group; and therefore the trouble of 'digging one out' is saved us. If we can afford to neglect units in the equator this means an immense simplification of our catalogue. But it must not be supposed that the units of the S group are simply units in the U group. They are units which are defined as *one or other* of the units of a U_0 of the U group. The importance of this distinction will appear if we pass in review the units in the S group corresponding to U_0s in a U_1 of the U group; say a ray of the pencil of U_1s, through O_1. As the U_0 under review passes towards the equator, the unit in the S group passes to the equator, and when the U_0 is *in* the equator the unit in the S group is indeterminate. But the moment the U_0 has passed through the equator, the unit in the S group reappears *as the antipodal in the U group* of the unit we were tracing before. Hence, though in one sense the unit of the S group corresponds to the unit in the U group, in another and more important sense it corresponds to the U_0. For *the equator is not a boundary* in the S group. And further, though we can say that one unit in a U_1 is between two others (in the same sense that we can in the U system), yet we can *not say that a U_1 is a boundary in a U_2*, any more than we can in the M system, or that a single unit is a boundary in a U_1. We can say that a U_1 in a U_2 is 'between' two units, but not that it divides the U_2 into two distinct parts. Again, though, as in the U system, we can say that one unit is between two others, in accordance with our convention, yet *we can always collate by projection any three units in a U_1 with any other three*; for whether we take the third unit or the antipodal of the third in the U system, we take in the S whichever of them is in the half of the U system we are using as a catalogue. B groups in the S system will appear in various forms. Either one or portions of both the antipodal groups may be within it. C groups would be particularly interesting to discuss, but I refrain. If parts of both antipodal groups are within the S group the distinction of 'within' and 'between' the groups may be maintained If not, we may say 'within' and 'without.' In either case they are boundary groups, while A groups are not. Two units always determine a U_1 unless both are in the equator. Two U_1s in a U_2 have always one unit common, and may have two, if they are in the equator. Two U_2s in a U_3 always have a U_1 common, but it may be a U_1 in the equator. I need not say more, as these consequences are sufficiently obvious and familiar to us all.

39. There remains one, or perhaps I should say two, more systems of cataloguing to be discussed. Instead of taking one half of the U group to catalogue a collection I might take less than half, namely, a portion defined by a boundary group in the U group. I might take either one of the portions within one of the antipodal groups, (I shall call this the H system) or the portion between the antipodal groups (I shall call this the I system). It will be seen that there is a very close logical relationship between the two systems, though for practical purposes one may be much more convenient than the other. The boundary in the U group would naturally, if not quite necessarily, be taken to be one of the C groups which we used in collating the pencil of $(U_1)s$ with origin O_1. Thus the H group would be symmetrical with respect to the origin, and the I group with respect to the section, of the first pencil of $(U_1)s$. Any U_1 would either lie wholly in the I group, or there would be only a single portion of it in the H group, since a U_1 can not have more than two $(U_0)s$ common with a C group. (If we were to take any other B group as boundary we might have the U_1 determined by two units in an H group divided into two distinct parts within the H group). The logical units in the H group would be single units of the U group; in the I group on the other hand we might take either units or U_0s, that is we might catalogue the section of the pencil on the M or U, or S systems, the only U_1s in the I system which have as yet been

catalogued being the rays of the first pencil. It is, however, not worth while for my present purpose to enquire further into the I system of cataloguing. I shall treat it merely as the correlative of the H system. It will be found that in most respects the peculiarities of the H system correspond with those of the S, and those of the I with those of the U. We may in fact regard an S group as a limiting case of an H group, when the I group has shrunk up to nothing more than the equator; and similarly an U group as a limiting case of an I group, when the H group has shrunk up to the origin.

40. If we pass in review the units of a U_1 in the H system, starting from an unit within the boundary, we shall sooner or later come to the boundary and have to stop. There is no unique relation between the terminal units in the U_1 as there is in the S system, so we can not in any sense look upon them as contiguous elements. I may mention that this and the other conclusions I am about to draw are not merely due to the way we are looking at the question, namely, considering the H group as part of a complete U group. They can be proved independently, from considerations affecting units in the H group itself alone.

In the H system, as in the S and U systems, we can say of one unit in a U_1 that it is 'between' two others. But more than this, we can say that (together with the terminal units, or one of them), it forms a 'boundary.' Similarly in a U_r, a U_{r-1}, (together with the boundary group), forms a boundary. It divides the whole group into two distinct parts.

We may have the whole of one of the groups of units in a B group of $(U_0)s$ within the H group, or we may have only a part of it, or we may have part of each of the two groups of units. If, however, parts of both groups are in the H group, there is no longer any essential distinction between the three parts into which the H group is divided. It is only by taking into consideration parts of the B group without the boundary that we can distinguish between the part 'between' and the parts 'within' the two groups.

Within the H group an A_1 may have an even number of units common with another A group, or none at all. And with a B group it may have an odd number common, for the remaining common units may be outside the H group.

We cannot always collate by projection any three units in a U_1 with any three, unless we are at liberty to take origins of projection outside the boundary of the H group. Nor can it be said that the order of units in two sections of the same pencil of $(U_1)s$ is identical, unless we neglect the boundary, for the boundary will not in general come in between corresponding units in the two sections. If we ignore units beyond the boundary, there may be no units to collate with certain of the units of the initial section.

The principle of duality applies in a yet more restricted manner than in the U and S systems. In a U_2 two U_1s do not always determine an unit or U_0 at all in the H and I systems; for they may determine one beyond the boundary, which would not count. There is, however, another kind of principle of duality which exists reciprocally between the H and I systems. It may be shown by the ordinary methods of projective geometry, that to each unit in the H group as 'pole' there corresponds a (U_{n-1}) wholly in the I group as 'polar.' If we allow ourselves to discuss the I system as an 'imaginary' complement of the H system, we get in this way an 'imaginary' principle of duality which may serve to obscure the anomalies of the H method of cataloguing.

41. It appears from this discussion that the five systems are not logically correlative. The most simple and logically perfect is the M system, in which the U_1 is closed, but no unique relation is postulated between any two of its units, or between any one and the group as a whole. The U and S systems come next in order of simplicity, and the anomalies in the one are in a sense complementary to those in the other. In the U system, like the M, the U_1 is a 'closed' group, whereas in the S system it is 'open.' In the U system we postulate an 'antipodal' relation between each unit and a certain other unit; in the S system this is only done in the case of units in 'the equator.' The H and I systems are also complementary to each other, as the S and U are, the U_1 in the H system being open, and in the I system sometimes open and sometimes closed.

But the *H*, *I* systems are also complementary in another sense. It is only by arbitrarily shutting our eyes to units 'beyond the boundary' that they can be considered separately at all; whereas in the *U* and *S* systems this is not the case. We may use abusive epithets towards units outside the pale, and call them 'imaginary,' but in any complete logical discussion of either the *H* or the *I* system they will intrude themselves, nevertheless.

42. In the course of the discussion so far I have been representing units by letters of the alphabet or numbers, and in certain cases I have assumed that these symbols themselves had a well understood order. But the order in question has never been more than a discrete order, the catalogue which in reality I have employed has been the time conception. And that is why I have not attempted to catalogue groups of more than the first order, except by collation of unique groups as pencils of the first order. The time conception is the first and best understood conception of a continuous group, and 'passing in review' the units in a group of the first order is, in effect, nothing but collating them with units of the time series, used as a catalogue. Until the group of numbers became 'well understood,' as a continuous group, it was not available as a catalogue for continuous groups; and accordingly in primitive attempts at cataloguing such groups it is not employed. I cannot say historically when it first occurred to anybody to employ it so; but the first attempt is popularly associated with the name of Descartes. He however did not regard the question from quite so abstract a point of view as I have done, and it will be worth while to go over the ground again in relation to my theory.

The advantage of cataloguing groups of the first order by means of numbers is that they constitute a group which we can 'carry in our heads,' so that we instinctively recognise its 'order' in the vulgar, discrete, sense of the word. But before we can apply it to the cataloguing of continuous groups we must understand its order in the extended meaning of the term. We must be able to 'project' the number series, that is, to apply such transformations to each of its units as will enable us to collate any three with any other three, but which shall then determine uniquely what fourth is collated with any given fourth.

43. It is evident that whatever transformation we apply must not alter the order of the units in the discrete sense, though it may be that not every possible transformation which fulfils this preliminary condition would in other respects be suitable. The most general transformation of this kind would in mathematical language be described as ' taking a one-valued function ' of each of the numbers. But a few preliminary considerations will enable us to divide the business of ' taking ' this function into stages, and so simplify the matter.

In the first place, if we call an unit a, in a group of the first order, '1'; b, '2'; and c, '3,' then, if we want to call b, ' 1 ' instead of a, we may call c, ' 2,' and a, ' 0.' This is obvious in the case of discrete groups. But further, it is generally obvious that we may always call any one unit by any number we please, by adding or subtracting from each number in the series an appropriate constant. The only units whose numbers will remain the same in spite of this treatment, will be the units we have called ∞ and $-\infty$.

Or else if we wish to call a ' 2 ' instead of ' 1,' we may do so by multiplying each unit in the series by 2, instead of adding one to each unit. In this way we may call any unit by any new number, except that the units numbered zero and $\pm \infty$ will remain unaltered. By these two methods we can change the numbers of any two units for any other two, and so we have only to determine in addition a 'function' to change the number of a third in an assigned manner.

44. To do this, let us proceed thus. Suppose the series of numbers to be represented by the algebraical symbols $\alpha, \beta, \gamma, \delta, \epsilon$, and so on ; the symbols being ' collated ' with the numbers, but not employed to catalogue them alphabetically.

First subtract a constant number, say α, from each. Then take a one-valued function F, which I may call the ' projective function,' of each of the differences. Next subtract again a constant number, this time $F(\beta - \alpha)$, from each. And lastly multiply the number so derived in each case by a constant number, h. The derived series of numbers is

$$h\{F(o) - F(\beta - \alpha)\} ; \ o ; \ h\{F(\gamma - \alpha) - F(\beta - \alpha)\} ;$$
$$h\{F(\delta - \alpha) - F(\beta - \alpha)\}, \text{ etc.}$$

I may call this process *projecting* the series of numbers, for it collates the units in the original series with the corresponding ones in the derived series.

Now project the series all over again, using a different function, F' instead of F, and a different constant, h' instead of h, if we prefer to do so; putting any three numbers $\alpha'\beta'\gamma'$ which it is desired arbitrarily to collate with $\alpha\ \beta\ \gamma$ in their places, respectively, in the process. They will be collated through the two derived series as an 'intermediate section,' if we can identify the numbers derived from them respectively.

The numbers derived from β, β' are already identical, both being zero. The numbers derived from γ, γ' can be made so by choosing the ratio $h : h'$ suitably. And whatever ratio is selected the numbers derived from α, α' will be identical *if $F(0)$ and $F'(0)$ are both infinite.* And now, if δ' is the number collated with δ, it is determined by the equation

$$\frac{F(\delta - \alpha) - F(\beta - \alpha)}{F(\gamma - \alpha) - F(\beta - \alpha)} = \frac{F'(\delta' - \alpha') - F'(\beta' - \alpha')}{F'(\gamma' - \alpha') - F'(\beta' - \alpha')}.$$

The only conditions which, so far, limit the functions F and F', are that they shall be one-valued, and that they shall be infinite for zero value of the argument. They need not necessarily be the same function.

This ratio, which I may call k, is the relation between four numbers, in a continuous series, which remains unaltered by projection. It defines, in fact, their 'order' in the continuous sense. With α', β', and γ' constantly collated with α, β, γ, any value of k determines two units δ and δ', which are collated together.

The particular projective function, F, to be adopted in any case, will depend on the system of cataloguing. But where we are dealing with rays of pencils of (U_1)s catalogued all on one principle, as in reversion, it will be the same for all.

45. Now suppose that the (continuous) order of the units $\alpha, \beta, \gamma, \delta$ is the same as that of $\alpha, \beta, \delta, \gamma$ (it cannot, of course, be the order in which I have written the symbols). That is, suppose we revert γ into δ with respect to (α, β) as pivot and equator. Then

$$\frac{F(\delta-\alpha)-F(\beta-\alpha)}{F(\gamma-\alpha)-F(\beta-\alpha)}=\frac{F(\gamma-\alpha)-F(\beta-\alpha)}{F(\delta-\alpha)-F(\beta-\alpha)},$$

$$\therefore \ \{F(\delta-\alpha)-F(\beta-\alpha)\}^2=\{F(\gamma-\alpha)-F(\beta-\alpha)\}^2.$$

Therefore either $\qquad F(\gamma-\alpha)=F(\delta-\alpha)$ ....(i),

or $\qquad F(\delta-\alpha)+F(\gamma-\alpha)=2F(\beta-\alpha)$ ....(ii).

Now equation (i) would represent a 'singular' reversion, for only three units enter into it; and singular reversion is only possible if we discriminate between antipodals. In this case therefore γ and δ must be antipodal, namely the units in the equator of the reversion, and equation (ii) would not be true unless β were identical with one or other of these units. In the alternative, that is if γ, δ are not antipodals in the U system, or in any case in any other system, reversion is represented by equation (ii). We may say that ordinary reversion is always represented by (ii), and that in the U system, where alone it is generally possible, singular reversion is represented by (i). The former case is given by $k=-1$, the latter by $k=+1$.

46. We may now investigate the series of numbers appropriate to the various systems of cataloguing. In every case it must be remembered that any one unit, or U_0, in a U_1, may be catalogued by any number. There is no reason why more than one number should not be assigned to each unit, or U_0, but to each number there must unambiguously be assigned one unit, or U_0, as the case may be.

In the M system there is no distinction between any one unit of thought (U_0) and another, and if we pass in review the whole group, starting from one numbered, say zero, we shall in the end come back to the same U_0; and this time suppose we number it s. If we pass the whole group in review again, when we return to the initial unit we shall number it $2s$, and so on. If we start with any other U_0, numbered α, when we come back to it it will be numbered $(\alpha+s)$, when we get to it a second time $(\alpha+2s)$, and so on. Hence $(\alpha+ns)$, where n is any positive or negative integer, must be collated with the same U_0 as α. If now we perform the process of projection upon the series of numbers so obtained, since $(\alpha+ns)$ represents the same U_0 as α, $F(\alpha+ns)$ must represent the same U_0 as $F(\alpha)$. And, moreover, the function of α plus anything but an integral multiple of s, must *not* represent the same U_0 as function of α. Hence the projective function, which in this case we may denote by M, must be a periodic function with the single period s. Either the tangent or cotangent would answer this condition, but since it must be infinite for zero value of the argument, it must be the latter. Hence

$$M(\sigma) = \cot \frac{\pi}{s} \sigma.$$

Similar considerations show that in the case of the U system the projective function must also be periodic. But in this case we have the anomaly that we can not always project three units on to any other three—there is an ambiguity as to whether it is in any case the unit we want or its antipodal, on to which we have projected a given unit. In the cataloguing by numbers the corresponding ambiguity must be that after projection we do not distinguish between a fourth unit and its antipodal. Hence in this case it may be that the projective functions for an unit and its antipodal are identical. This we find to be the case. For if t be the whole cycle, so that $(\alpha + nt)$ is collated with the same unit as α, and if γ be the antipodal of α, as we pass in review the units from γ to $(\alpha + t)$ the antipodal of the unit under review passes from α to γ. Hence

$$\alpha + t - \gamma = \gamma - \alpha \text{ or } \gamma - \alpha = \frac{t}{2}.$$

And consequently the equation representing singular reversion shows us that

$$U(\theta) = U\left(\theta + \frac{t}{2}\right)$$

and as in the case of the M system we deduce

$$U(\theta) = \cot \frac{2\pi}{t} \theta.$$

In the S system the only two antipodal units in a U_1 are those in the equator. We must therefore be able to revert one of these into the other with any pivot, by singular reversion. That is, if we call them γ, δ, we must have

$$S(\gamma - \alpha) = S(\delta - \alpha)$$

for all values of α. Further, there must be no units in the series of numbers beyond γ and δ, in either direction. Hence

γ and δ must respectively be positive and negative infinity; and as in addition $S(o) = \infty$, we may put

$$S(\rho) = \frac{r}{\rho}$$

where r is any constant number.

In the *HI* systems, as in the case of the *S*, we can have no units in the series of numbers beyond those collated with the boundary—at least any numbers which there are must for some reason be held not to count. The easiest way of finding the appropriate projective function is to deduce it from that of the *S* system. The boundary in either case is a *C* group whose units, on the *S*.system, are all collated together. Hence on the *S* system they all have the same number. As in the *S* function we have an arbitrary constant at our disposal it does not matter what number we choose, so for simplicity I take unity. Hence, if ϕ be the number collated on the *H* system, ψ that on the *I* system, with the unit collated with ρ on the *S* system, we must have ϕ and ψ both infinite when ρ is unity, and when ρ is zero ϕ must be zero, when ρ is infinite ψ must be zero. These conditions are fulfilled by putting

$$\phi = p \log \frac{1+\rho}{1-\rho}, \qquad \psi = q \log \frac{\rho+1}{\rho-1},$$

$$\rho = \tanh \frac{2\phi}{p}, \qquad \rho = \coth \frac{2\psi}{q},$$

$$H(\phi) = \coth \frac{2}{p}\phi, \qquad I(\psi) = \tanh \frac{2}{q}\psi.$$

Now in reality either of these systems of numbers catalogue the whole of the *U* group, for the *I* part is catalogued by the imaginary numbers on the *H* system, and the *H* part by imaginary numbers on the *I*. For symbolic purposes therefore there can be no possible objection to the *HI* systems. It is only when we try to give them real import that any difficulty arises.

To transform from the *S* to the *U* or *M* systems as we have done for the *H* and *I* it is clearly only necessary to write

$$\rho = \tan \frac{2\pi}{t}\theta = \tan \frac{\pi}{s}\sigma.$$

47. On any system of cataloguing it is convenient, though

not necessary, to assign the numbers so that the fundamental
reversions in the process of cataloguing shall be purely arithmetical reversions, that is, that they shall be effected with
reference to zero as a pivot by a mere change of sign. By this
convention if β be the unit collated with the equator in the U
system, and $\beta - \dfrac{t}{2}$ that collated with its antipodal, also in the
equator,

$$\beta = -\beta + \frac{t}{2}, \quad \therefore \; \beta = \frac{t}{4}.$$

And if γ be the unit collated with P_1, or p_1, and we move the
zero of the series of numbers to P_1 by substracting γ from each,
we must have

$$\beta - \gamma = \gamma, \quad \therefore \; \gamma = \frac{\beta}{2} = \frac{t}{8}.$$

In the M system, by similar reasoning, the $(U_0)s$ on the
equator will be collated with $\pm \dfrac{s}{2}$ those in the C_1 with $\pm \dfrac{s}{4}$. In
the S system those in the equator, as we have already seen, are
collated with $\pm \infty$, and those in the C_1 by this convention will
also be collated with $\pm \infty$, that is, they will not be distinguished
from the units in the equator. In the H system the units in
the equator will be collated with the 'imaginary' number $i\pi$,
which will be collated with the origin in the I system. From
the point of view of pure logic 'imaginary' numbers are just
as good as 'real' ones; we might use them just as effectively
for cataloguing a library (for the use of mathematicians only).
To the mathematician who employs numbers to catalogue a
collection on the H system, the I system will always be present
in imagination; he will call it 'imaginary.' It is only when we
try to give real import to the term that 'imaginary' comes to
mean 'not real.'

48. I have now laid the foundations of the science of the
Cataloguing of Continuous Groups; but before proceeding to
consider its applications I wish to impress upon you very
strongly that it is not geometry at all. It is not only not
metrical, and not spatial, but it is not even necessarily numerical. I have shown how groups may be catalogued by the aid

of the number series, but I have also shown how they may be catalogued without it. Geometry, and also Arithmetic, are developments in special directions, or rather special applications, of a far more general or abstract theory, which Mr. Kempe calls the theory of Mathematical Form, and which includes the theory of cataloguing of both continuous and discrete groups, with or without the aid of numbers. It is almost inevitable in thinking of the theory of continuous groups, at least of those of lower order than the fourth, that we should attach more or less spatial import to the terms. If, however, I have succeeded in my exposition, it will be clear that this spatial import is not necessary to the logical validity of the arguments employed. The 'experience' upon which they depend is not experience of 'externality,' except in the somewhat ambiguous sense in which Time is spoken of as a 'form of externality.'

49. I may further point out that I do not, of course, claim any originality for the propositions I have proved, considered as propositions in 'projective geometry.' The object of the exposition has rather been to show that they are not essentially geometrical propositions at all. I accept Mr. Russell's authority for saying the philosophical principles of Projective Geometry required further elucidation, though I cannot accept his exposition as a further elucidation of them.

PART III

But if it is clearly realised that the science of the cataloguing of continuous groups is not *per se* geometry, I may go on to show how it is related to the latter science. And this can, I think, be done without prejudging the question whether space 'is' or 'is not' Euclidian.

Mr. Russell, after revelling in contradictions in a manner of which only a modern philosopher could fail to see the humorous side, arrives at the conclusion that we must "give every geometrical proposition a certain reference to matter in general," and thus distinguish a point from a position. He might have found this conclusion much more simply expressed in my *Foundations of Geometry.** I gave as the implicit definition of position, the two assertions " (*a*) A position may be conceived to be indicated by a portion of matter, called a point, which is so small that for the purpose in hand variations of position within it may be neglected ; (*b*) But a position is not the same thing as a point, for a point may be conceived to move, that is to change its position, whereas to talk of a position as moving, is a contradiction in terms."

I carried out the same idea by distinguishing between a (material) line and a 'path,' between an angle, the concrete measure, and an 'inclination,' the abstract relation. I propose to adopt similar distinctions here ; for I could not, like a modern philosopher, go on, after admitting that my definitions contained "a fundamental contradiction." I must however point out that I do not, and did not in my former treatise, use the term 'matter' in an objective sense, any more than, I presume, Mr. Russell does in the passage I have quoted. The geometry I was discussing then, and which I am discussing now, is purely subjective, or ideal, geometry.

It is obvious that the group of positions in a line is a group of the first order, which we may, and habitually do, collate with

* **Deighton Bell and Co., 1891.**

the time series directly. I imagine myself as a material point passing along the line. The line may be conceived as open or closed, and taken by itself I might, if it is closed, catalogue its units on the M or U systems, or if open, on the H or S. So far it is obvious that the choice of system is arbitrary. But further, if it is regarded solely by itself, there is no logical reason for regarding it as open rather than closed. The distinction lies in the fact that, in conceiving the line 'in space,' I have already catalogued its points by means of my conception of space as a catalogue. This would prevent my passing in review from one end of a terminated line to the other ; such a line must therefore be catalogued on the H system, if it is to be catalogued by itself. The conception of an unterminated but open line, if we do conceive such, would, as we shall see presently, indicate that we had adopted the S system of cataloguing in our conception of space. The conception of a *straight* line has undoubtedly the property that 'in general' it is determined by two points, that is, we conceive the positions in a straight line as a U_1 of positions, and those positions are the logical units of our catalogue ; though, if it is an M catalogue, they are what I have called $(U_0)s$. Further, it will be easily admitted that the group of positions in a surface is a group of the second order, and the group of positions in a volume one of the third order. In a closed or bounded surface, a closed line, or one terminating in the boundary, divides it into two distinct parts, and in a bounded volume a surface terminated in the boundary divides it into two parts. And as we conceive three points not in a straight line as in general determining a plane uniquely, the positions in a plane form a U_2. As, however, we do not conceive any positions outside a group of the third order, as this group is our whole collection of positions, there is *prima facie* no point in calling our whole space a U_3 rather than a G_3.

Now, how do we 'collate' our conceptual space ? We do so by ' the method of superposition.' That is, we pass in review the series of positions occupied by points in a line or surface, which is supposed, in virtue of the axiom of congruence, "*Es existiren in sich feste Körper*," to remain identical with itself. We do not, however, conceive the points to pass along straight lines which meet in a point—we do not collate by projection. The line or figure in superposition moves, that is, changes its positions, by either 'rotation' or 'translation.' But if our space is catalogued on the U or M systems these two movements are identical in nature, if on the S system the only distinction is that in translation the axis of the rotation is in the equator, and on the H or I system that it is beyond the boundary, so that translation is an 'imaginary' rotation. We may say, therefore, that superposition is effected by one or more rotations, and in one rotation the axis of rotation 'remains fixed,' *i.e.*, is always collated with itself, and the positions through which the other points of the figure pass are C_is. In fact, collation by superposition is collation by reversion. The locus of each point is a group of positions of the first order, such that no three positions in it are in a straight line. The further properties of a circle—which distinguish it from other conics—are due to the assumption that something more than the order of the points in a line remains identical as we move it about. Of this I shall say more presently.

Now our conceptions of space are essentially egoistic. *I* am am the hub of my own universe; and it is in reality always in relation to a position conceived to be occupied by myself that I conceive space. And with reference to that position I conceive it to be perfectly symmetrical. This leads at once to the rejection of the I system as incompatible with our conception of space; for from that point of view, as we have seen, space would be regarded as symmetrical with respect to a plane, but not with respect to a point. We look at space primarily from the polar, not the cartesian, point of view. Now from the polar point of view we may, if we please, look upon one of the C groups, the locus of P_1 say, as the section, instead of the equator. That is to say, we may catalogue the rays of the pencil of (U_1)s by the units or (U_0)s in them in this C_1. In the case of space in the same way we may catalogue straight lines radiating into space in all directions from any origin by the positions of points in a sphere with that origin as centre. In this way we may obtain at once a catalogue of the rays, without begging the question whether the rays themselves are catalogued on the M, U, S, or H system. It is clear that the two points on one ray are related as antipodals, and that the catalogue of the 'celestial sphere' is on the U system.

The logical consequences which would flow from the adoption of the U, S and H systems have been sufficiently worked out by the meta-geometers, and they have further been illustrated by actually cataloguing groups of points on various surfaces in space, with or without the assumption of an axiom equivalent to the axiom of congruence. In the illustrations of two dimensional spaces by surfaces of constant curvature such an axiom is assumed, though in the case of surfaces of negative curvature it is not assumed in quite the natural sense. In illustrations of three dimensional hyperbolic spaces the axiom is, however, frankly abandoned. Nevertheless, Mr. Russell regards these illustrations as necessary to " prove the possibility " of U and H spaces, and I shall therefore give a similar illus-tration of an M space; after which Mr. Russell must hold it to be just as " possible " as the others.

Mr. Russell gives an example of what he holds to constitute an objection to M geometry, the geometry of Klein's 'elliptic space.' I will reproduce what he says in a form slightly altered to suit my present purpose. Suppose I take a triangle, one corner of which is one yard in front of me, one corner one yard to the right, and the third one yard to the left; and I move this triangle, with the first corner and the middle point of the opposite side remaining in a fixed straight line, 'round the universe'; so that the corners to the right and left of me return to where they were before. If 'the universe' were a pencil of rays grouped on the M system, the result would be that the first time the triangle came round its vertex would be *behind* me, and I should have to move it round a second time in order to get it in front of me again. Mr. Russell says this "*seems* like an action of empty space;" and though I can not quite follow his reason for rejecting this appearance, I quite agree with him that it is not so. After what I have said about the M system the analogy I am about to propose is obvious. Instead of the triangle as we have conceived it, push the triangle a couple of yards upwards into the air; and for each corner substitute the

straight line determined by my position and the corner; for each
side substitute the plane determined by me and the side. Both
lines and planes so substituted are to be continued upwards and
downwards through my position, indefinitely. Now in order
that the lines representing the corners of the base may return
to their initial situation, it is only necessary that the plane containing them should be related through two right angles. But
when this is done, the line representing the vertex will *not* have
returned to its initial situation. To bring all three into congruence at once, we must rotate through *four* right angles.

And generally, just as the points on a sphere illustrate the *U*
system, so the diameters of the sphere illustrate the *M* system
of cataloguing.

It might thoughtlessly be objected to this illustration that
after all it was not a genuine illustration of *M* space, since lines
are *not* points. Now if the illustration is to be regarded as
showing that it is possible that we do catalogue space on the *M*
system, this objection is perfectly valid. Only, there are similar
objections to all the illustrations which have been given of *U*
and *H* spaces. The great circles on a sphere are *not* straight
lines, any more than the diameters of it are points. The
geodesics of a surface of constant negative curvature are *not*
rigid lines, though they are rigid against bending in the surface.
We have, in both cases, substituted new conceptions and called
them by old names. The same is even more obvious when, for
example, Poincaré substitutes for 'distance' the logarithm of a
certain anharmonic ratio. All these illustrations illustrate the
logical processes involved in meta-geometry, but they do *not* in
the least degree illustrate the psychological processes which
would be required to conceive space as non-Euclidian. The
question whether the space we conceive is Euclidian or not, is
not a logical question at all, but a psychological one. It can be
determined with certainty; not indeed with the verbal certainty
of formal logic, but with the absolute certainty of direct intuition.
I have already attacked this question of subjective geometry in
my book on the *Foundations of Geometry*, and from the point
of view which I there took up I have nothing particular to add.
I showed there, upon what I still believe to be perfectly sound

logical and psychological principles, that our subjective geometry is Euclidian, and no other. It may, however, throw some light upon what I still consider to be psychologically the fundamental elements of our space-conception (position, and direction, conceived as independent of each other), to bring the views I then expressed into relation with the theory of the cataloguing of continuous groups.

I have already pointed out that in the latter theory collation by reversion corresponds logically to the method of superposition, but that psychologically the axiom of mobility, or congruence, implies more than this. The theory of cataloguing by reversion would allow us to collate together points on an ellipse just as well as points on a circle, in the rays of a pencil of straight lines of the first order; it would not even be necessary that the origin should be the centre of the ellipse. The conception of distance as a relation between the positions of points in a straight line, which is unaltered when the straight line is moved so that the points occupy new positions, is psychologically a *different* conception from that of the order of the points. And consequently, if we conceive this sort of relation between the positions of space, space is no longer *any* kind of continuous group. I have already shown that it follows at once, that it is not an I group. If this were the only new psychological relation assumed between the positions of 'empty space,' it would remain indeterminate whether the conception was that of an M, U, S, or H group. But as a matter of fact we conceive another relation between two positions besides distance, namely direction. It is commonly said that this conception is not independent of the other, that no meaning can be attached to the statement that the direction from A to B is the same as the direction from C to D, unless all four points are in one straight line; or else, unless we assume space to be Euclidian. This argument is logically sound, but psychologically it is putting the cart before the horse. We *do* attach a meaning to the statement, and that proves that our space-conception *is* Euclidian, and no other. I need not repeat what I have said elsewhere on this point, but will consider more particularly some of the psychological absurdities to which the statement that our conceptual space is non-Euclidian would lead us.   I must first, however, point out one important philosophical consequence of the psychological import of the two conceptions of distance and direction, as relations between two positions.   It is that these elementary conceptions define not only the form in which we conceive particular figures, but *also* the form in which we conceive 'empty space' as a whole.   They are independent, not only of each other, but of what Mr. Russell calls the axiom of dimensions.   Just as we might conceivably have a different conception of direction, such, for example, as would be applicable to an *H* or *U* group, so we might conceivably have a different conception of the number of dimensions of our 'empty space,' without our other elementary space-conceptions being changed.   We might in this way conceive space as four-dimensional, and I believe that, psychologically, this change would be much more easily effected than any other.   In my *Foundations of Geometry*, I even went so far as to express the opinion that we might come to conceive a space of four dimensions in the course of a single life-time.   But I do not think it would be possible for most men, probably not for any, to obtain a new conception of direction in the same sense that they have the Euclidian conception; and philosophers in general seem to be agreed (though they do not express their agreement in this form), that it would be impossible to alter our conception of distance.   It is admitted that meta-geometrical spaces are 'Euclidian in their smallest parts,' and that therefore the Euclidian conception of space would be applicable to small figures, even if not to large ones.   I am convinced that if we should find the Euclidian conception which we have of space inapplicable to large objective figures, the easiest way out of the difficulty would be to try if we could not conceive material space as a boundary in an 'empty' Euclidian space of four dimensions.   However, this sort of speculation has little practical value, and I return to more important considerations.

The most obvious of them is found in the answer to the question—Do we conceive a straight line to be open, or closed ?   The answer to this is indeed so obvious that very little attention has been given to Riemann's spherical space, and still less to elliptical, or monistic, in comparison with 'hyperbolic' space.

It was perceived that, logically, spherical space was as valid as
hyperbolic—that the one was just as *a priori* possible as the
other. That, however, spherical space has never been considered
to be a practical possibility—that, for example, no astronomer,
as far as I am aware, has attempted to measure large triangles
for *excess*, but only for *defect*, is due to an instinctive recognition
of the importance of psychological considerations, combined with
an erroneous ascription to them of objective validity. It has
indeed been urged that we really conceive only a small part of
a straight line, just as, objectively, we can measure only a small
part of it; and that beyond the part we conceive it might return
and close on itself. The answer, of course, is that the straight
line in our conceptional space is what we conceive it to be, and
nothing else. If I imagine a straight line which is not closed,
to be produced and then to become closed, I have altered my
original conception, and the produced line is no longer 'straight'
in the sense in which it was before. Moreover, I conceive shape
to be independent of size, and to say a closed line is infinitely
great does not alter my conception of its shape in the least.
As, however, this may be held to beg the question at issue, for
it has been noted that Euclid's disputed axioms might be
deduced from the postulate that shape is independent of size,
I will waive this point, and examine some of the further
consequences of assuming that a straight line may be closed.

If I conceive two straight lines in a plane, each of them
perpendicular to a third, it may be shown by the method of
superposition that the figure is symmetrical with respect to the
third line; that is, that the other two must intersect in *both*
directions from the third line, or in *neither*. Now, as far as I
conceive the two straight lines, they certainly do not intersect—
I should not call them 'straight' if they did; and to say that
the straight lines which I conceive intersect somewhere else,
where I do not conceive them as intersecting, is to talk nonsense.
Subjectively, therefore, I hold that Euclid was perfectly correct
in saying that parallel straight lines do not intersect at all. But
as long as we are admittedly only talking symbolically it is quite
permissible to talk of parallel straight lines as intersecting 'at
infinity'; that being a locality which it is understood that we

do not conceive at all. In symbolic reasoning ∞ is no better, and no worse, than any other symbol, if it is consistently used. But as soon as we attempt to give any real subjective import to such an expression as 'an infinite distance' or 'points at infinity' we find we are unable to do so; they remain purely symbolic terms. This might be a serious drawback to our space-conception were it not compensated for by the fact that we conceive shape to be independent of size, and that consequently we may discuss what happens at home without troubling ourselves about what might or might not be said to happen 'at infinity.' It is as easy to conceive the motion of the solar system through space, as that of a vortex ring constituting an atom of matter, for we may discuss the shape of figures independently of their sizes. But we can not in any true sense be said to *conceive* 'points at infinity'; and though to talk of such points is not logically false, "*non est geometria.*" Geometry used to be regarded as the most perfect of sciences, because its terms could all be 'well understood'; and old fashioned geometers took care so to exhibit their proofs that this should always remain so. The modern habit of employing purely symbolic arguments, and still calling the science geometry, is calculated to lead to serious logical misconceptions—if indeed it does not indicate that they exist already. I know some people who believe that the 'asymptotes of an ellipse,' or the 'circular points at infinity' really exist somewhere, only that our drawing boards are not quite big enough to hold them. It ought to be clearly understood, that when we begin to talk about what happens 'at infinity' we have left the ground of pure geometry and have launched forth into abstract analysis.

However, the way of abstract analysis is a good way of acquiring knowledge, if it is properly employed, and if we are careful not to use our terms inconsistently. It is perfectly legitimate to say, symbolically, that two straight lines with a common perpendicular intersect 'at infinity'—if we say that they intersect in *both* directions from the common perpendicular. But it is not permissible to speak of them as intersecting on a straight line, or plane, 'at infinity,' as is so often done. For, either we must say that the intersections in the two directions from the common perpendicular are one and the same point 'at infinity,' or that they are two different points. If we say they are the same point, then the straight lines are regarded as limiting cases of closed curves, and if we transfer our attention to the point 'at infinity' where they intersect, we see at once that we are regarding space as a limiting case of one of Klein's elliptic spaces, or as I should put it, that we are cataloguing positions on the M system, with infinity as the period of our projective function. In this case the locus of the intersections of straight lines in a plane which have a common perpendicular within a finite distance of the origin, would be the infinite plane itself, and not a straight line in that plane, 'at infinity.' Similarly the locus of the intersections of such lines in space would be infinite space itself, and not a plane 'at infinity.' If on the other hand we say that two straight lines which have a common perpendicular within a cognisible distance of us, *i.e.*, the parallel straight lines of Euclidian geometry, intersect in two distinct 'points at infinity,' these two points are uniquely related as antipodal points, and their locus is, not a plane in the Euclidian sense of the term, for the points in a plane have no antipodal points, but what may appropriately be called a great sphere 'at infinity.' To use the same term for this great sphere and an Euclidian plane leads to the 'antinomies.' In this way of looking at it we regard Euclidian space as the limit, not of a Reimann's or spherical space, but of what I have called an S group of the third order; and it has this great advantage—that it does not involve saying that straight lines are closed lines, and therefore it constitutes the best symbolic representation of our space-conception. We do as a matter of fact conceive, not a boundary to our space, but a 'celestial sphere' ever so far off, beyond which we do not trouble ourselves to conceive anything. The points in which we conceive any straight line to reach this celestial great sphere are antipodal, and not identical, points. We can not, as I have said, be truly said to conceive this great sphere, but we can, quite naturally and easily, talk about it; and so avoid certain latent antinomies which surely we must all have suspected in the 'plane at infinity.'

It is worth while in this connection to point out an arithmetical error which is sometimes made, from a fancied geometrical analogy. It is sometimes thought that positive and negative infinity are arithmetically indistinguishable. We can not indeed say that the difference of two infinites is zero, unless we know by what process the limit was reached in each case. But even if we distinguish the two infinities by suffixes and write

$$+\infty_1 = -\infty_2$$

we can add to both sides of the equation identically the same infinity as the second, and so obtain, unambiguously, the result—

$$+\infty_1 + \infty_2 = 0,$$

that is, the sum of two positive quantities equated to zero; which is a contradiction in terms.

We may of course *collate* $+\infty$ and $-\infty$ with the same unit in a group of the first order, just as well as we may collate $+\pi$ and $-\pi$. But this is arithmetically equivalent only to equating a function of infinity with the same function of minus infinity, and it only shows that the function is a periodic one. The geometrical equivalent of this is the fact that I have already noted, that if we say two straight lines which have a common perpendicular intersect in one point only at infinity, then the straight line is a limiting case of a closed curve. It is of course only if we regard the straight line as open that the distance measured along it between two positions is unambiguously determined, or that it can be unambiguously represented by the difference of the numbers collated with the positions. It is therefore only in this case that we can regard the positions and numbers as *identified*, and not merely *collated*.

In my *Foundations of Geometry*, Part III., I have shown that if we take the axiom of mobility in its natural sense, that is, that neither the size nor the shape of geometrical figures depends on their position in space, no mere extension of the number of dimensions would allow us to conceive space as an hyperbolic (though we might conceive it as a spherical) boundary group in a higher group. The 'possibility' of hyperbolic space in this sense depends upon our ignoring changes of shape which do not affect 'sizes,' as determined solely by measurements in space. Though this view is of course logically permissible, I can not admit that it represents our actual conception of 'free mobility.' I, however, also quoted a 'proof of Euclid's XIth Axiom,' by a M. Vincent, which deserves more notice than it has hitherto received. I will briefly reproduce it here. I have already shown that we must say that two straight lines in a plane which have a common perpendicular either intersect each other on *both* sides of that perpendicular, or on *neither*. The proof is based on the latter assumption, and proceeds as follows. Two straight lines at right angles in a plane divide the whole plane into four equal quadrants. If we successively cut off from one of these quadrants infinite 'half-bands,' by straight lines through equi-distant points along one of the arms of the quadrant, and at right angles to it, then these half-bands can be shown by the method of superposition to be all equal to each other. But the remaining portion of the quadrant can be shown, by superposition, to be still equal to the original quadrant. Hence, we must say, that the area of a quadrant is infinitely great in comparison with that of any half-band, or of any finite number of half-bands. But a sector containing any finite angle is numerically comparable with a quadrant, and therefore infinitely greater than any half-band. Hence, if from the vertex of the original quadrant we draw a straight line making any finite angle with the perpendicular line, it must intersect the opposite side of any half-band.

Now the only objection I have ever heard urged against this proof is to the effect that it is a kind of sacrilege to talk about infinities in this way. No doubt it is very dangerous for incompetent persons to have anything to do with infinities, as it would be for anyone but a trained metaphysician to attempt any further determination of the absolute. But trained mathematicians are accustomed to deal with infinities every day, and are quite competent to determine whether this argument is sound or not. There need, however, be no doubt about the question, for it is easily shown that it is not really inconsistent with Lobatchewsky's geometry at all. It depends on the assumption, which is psychologically justifiable, that space is 'unbounded'; and meta-geometers admit that hyperbolic space

is *not* unbounded.   The importance of M. Vincent's proof is
that it brings home to us the fact that this peculiarity of hyperbolic space has a psychological, as well as a logical, significance.
Lobatchewsky may push his 'boundary' to infinity, but he cannot thereby push it out of sight altogether; its influence is still
felt at home.   We cannot regard space as hyperbolic, and
continue to assert the axiom of mobility in its ordinary sense
(even if we agree to neglect alterations of shape, such as occur
in figures moved about on a surface of constant negative
curvature, merely because they are not accompanied by alterations in the dimensions of the figures as measured along geodesics
of space).

There is one last point, before I leave the subject of conceptual
space, which I particularly commend to the consideration of
Mr. Russell.   It is that neither the principles of projective
geometry, nor the principle of duality, as enunciated by him,
are true without reservations of *any one* of the systems of
geometry which he regards as 'possible.'   They, strictly speaking,
apply only to the M system, to Klein's elliptic space.   In the
S system, or in Euclidian geometry, the reservations are so
slight that they may perhaps fairly be considered as merely
verbal.   But still there are exceptions, though it will be
sufficient to point out the more important anomalies in the
U and H systems, and leave those in the S to be inferred
from them.   In the U system it is *not true* that any three
units in a U_1 can be collated with any other three, or in
geometrical language, that any three collinear points can be
projected on to any other three.   We can, in general, only
project two on to two, and the third on to the third or its antipodal.   And in the H system it is *not true* that the order of
units in a U_1 is unaltered by projection; unless we take into
account units beyond the boundary.   That is, if we project
points in one 'straight line' in hyperbolic space on to another
straight line, there may, in the second, be no points at all
corresponding to certain points in the first.   It cannot therefore
be said that the second section is 'projectively indistinguishable'
from the first, unless we choose to recognise 'imaginary' points
in the second section.   But, if we do that, the *raison d'etre* of

hyperbolic geometry is removed. We are only calling things by different names, and not really conceiving anything different from Euclidian space at all.

I submit that the fact that Mr. Russell has overlooked these exceptions to the applicability of projective geometry to meta-geometrical spaces, is due to the fact that he never really conceived them as such ; that, in fact, his argument, so far as it is valid, is purely symbolic; and, in so far as it is invalid, it is so because it is tainted with spatial, and more particularly with Euclidian-spatial, prejudices.

I may mention, for the benefit of those who have not made a study of projective geometry, that the above is not the only geometrical application of the theory of continuous groups. In order that the positions of points in them may be regarded as forming $(U_1)s$ it is not necessary that lines should be 'straight,' as long as they are drawn according to some rule which enables us to consider space as a pencil of such lines with any origin, the lines being in any case determined uniquely by the origin of the pencil and one point in its section. The familiar illustration is the system of geodesics on a surface of constant curvature. The geodesics on any other surface would not do, for here, although any particular geodesics drawn might not have more than one or two common points, yet two points would not uniquely determine any one of them. The difficulty in the case of non-congruent surfaces is that we have no general rule by which unique groups of positions can be determined. But it may easily be shown that it is theoretically always possible to determine various systems of them. If we stretch an elastic membrane tightly over any congruent surface, say a plane or a sphere, and draw on it projective diagrams, in straight lines or great circles, and then remove the membrane and stretch it again over any other surface whatever, say an ellipsoid or an egg, the projective properties of the figures remain, and the lines, or as many of them as have been drawn, are unique boundaries. Similarly we might imagine a three-dimensional diagram, drawn on a lump of elastic jelly, and we might then screw up the jelly, or stretch or compress it, into any shape whatever, so long as we did not tear it anywhere, and our projective diagram would lose none of its properties, for the *order* of the points would be unaltered. It would of course be easy, without making use of an elastic jelly, to draw such projective diagrams 'free hand,' provided only that no two lines in the diagram had more than one common point, and that the diagram was only to contain a finite number of lines. The necessity for some *rule*, by which to draw the lines, would only arise if we wished to draw an infinite number, that is a continuous series, of them. In the case of surfaces of constant curvature we have such a rule, if we make the lines all geodesics. But it would be an error to suppose that the theory of continuous groups applies only to surfaces of constant curvature, or only to geodesic lines upon them.

As in our space-conception we have a well-understood example of a continuous group of the third order, catalogued by unique boundaries, we are able to use it as a catalogue for other continuous groups of the second and third order, and so to arrive at an intuitive perception of the relations among units of such groups. This is what we do when we assist our imagination by 'graphic' representations of statistical and other results. In the same way, if we admit the objective validity of our space conception, we may employ 'graphic' methods to solve statistical and other problems which involve two or more variables, whether these variables are in themselves spatial quantities or not. It is worthy of note that the word 'graphic' has acquired almost the same connotation as 'well-understood,' simply because we regard our space conception, not without truth, as 'well-understood.'

The questions which remain to be discussed are these—Can we consistently catalogue the positions of material points in the way I have described, collating them on the assumption that there is an invariable relation of distance between the positions of any two points in a rigid body, and that there is a relation of direction between any two positions? and if not, would it be necessary to alter our space theory, or might not the adjustment be made more simply by altering some of our physical hypotheses?

To these questions I now proceed, but if ever I use ordinary language, and say that material space 'is' or 'is not' Euclidian it must be understood that I do not intend thereby to institute any comparison between material and conceptual space. To perceive a material body is not to compare it with our conception of it, nor is it only to believe that it exists; but it is to intuitively catalogue its points by means of our space-conception, and to believe that method of cataloguing will be satisfactory. To say the material body occupied a material space which was 'like' our conceptual space, would lead to 'antinomies,' and therefore I avoid doing so, whenever I do not forget myself.

Now the fact that we have an instinctive conception of space, and that geometry is psychologically distinguishable from such hypothetical theories as, for example, the wave theory of light, proves that for the ordinary purposes of every day life the ordinary theory of Euclidian geometry is, and has for generations been, good enough. But more than this; since the most far-reaching and exact calculations of geodesy and astronomy have been made on the Euclidian hypothesis, and have turned out satisfactorily, the theory has already been tested more severely than probably any other; and from the point of view of Natural Science there is nothing more to be done than to accept it fully and frankly.

But from the point of view of philosophy it is not so much the *fact* of the certainty which is of interest, as the epistemological grounds upon which it is based. I wish to show that its certainty does not differ in *kind*, however much it may differ in degree, from that of any other hypothesis of physical science.

Mr. Russell makes a great point of the superior certainty of the 'axiom of dimensions' over that of, say, the law of gravitation. The supposed superiority is however based on a misconception. The certainty that conceptual space has only three dimensions depends of course on utterly different grounds from that of the law of gravitation, but the certainty that our three-dimensional conception is objectively sufficient, is only the same in kind, even if superior in degree to that of any physical 'law.' Astronomers used anxiously to watch any new comet, to see whether it 'obeyed the law of gravitation' or not; and if they have now given up doing so, it is because they find it easier to explain any apparent discrepancies by hypotheses about meteor streams, and what not. In precisely the same way, Zöllner anxiously watched the medium Slade, and came to the conclusion that his strips of hide did not 'obey the law of three dimensions,' though perhaps other people might find simpler hypotheses to account for the results. Zöllner was however logically justified in assuming that the fact that a three-dimensional space hypothesis had been found sufficient to explain all previously observed phenomena, does not prove that it will account for the next phenomenon to be observed, any more than the fact that the law of gravitation explains the movements of the planets, proves that it will also explain the movements of the next new comet. There is, moreover, no sense in saying that material space ' is ' or ' is not ' three-dimensional, rather than *n* dimensional. All we can say is that the assumption of three independent directions has hitherto sufficed to catalogue points in space satisfactorily, and that you may safely ' bet your bottom dollar ' that it will continue to do so.

Besides the ' axiom of dimensions,' Mr. Russell recognises two others, which he calls the ' axiom of mobility ' and the ' axiom of distance,' and he compares the latter with the ' axiom of the straight line ' in projective geometry. I cannot however help thinking this classification implies some confusion of thought; for it is the axiom of mobility which he says gives us our criterion of spatial equality, and therefore our measure of distance, and which moreover serves to define a straight line in the sense in which that term is used in projective geometry. The term ' axiom of mobility ' seems to me unfortunate in another way—the axiom is not to the effect that it does not require force to move figures about in space; that would be a dynamical, not a geometrical, axiom. It may be easy or difficult to move figures in space; all that the geometrical axiom asserts is that when moved they do not, *ipso facto*, change their dimensions. But whatever we call the axiom, the effect of it is to enable us not only to compare objective distances, but to construct objective straight lines and planes, independently of any physical hypotheses about the nature of light. The assumption of this ' axiom ' is equivalent, in my theory of cataloguing of groups, to the two assumptions (1) that we can catalogue positions of material points as a continuous group by means of unique boundaries (straight lines and planes), and (2) that we can collate the positions by reversion in the way I have described above, by assuming an invariable relation of 'distance' between the positions of any two points in a rigid body. This second assumption, therefore, defines our method of cataloguing as either M, U, S, or H. The further axiom or hypothesis required to define 'material space as Euclidian' may fitly be called the hypothesis of direction, namely that there is a relation of *direction* as well as of *distance* between the positions of any two points of a material body; or we may call it the axiom of the independence of shape and size. It is this latter hypothesis, or axiom, which meta-geometers have seen fit to doubt, but it is a logical error to suppose that no physical experiments could throw a doubt upon the axiom of congruence, or that it can be known, independently of experience, that a space hypothesis built upon that axiom *must* fit the facts. It may be that under any conceivable circumstances it would be simpler to reconcile our explanations of the facts by altering some other hypothesis, rather than this one, but this is a matter of practical convenience, not of logical necessity.

I can best show that this is so by proposing a single experiment to test the rival theories of objective geometry, which will at the same time test the axiom of congruence. Mr. Russell has pointed out that even if we found that large stellar triangles had a measurable 'defect,' we should probably attempt to explain it by modifying our hypotheses as to the propagation of light through interstellar space. It would require a much more fundamental re-construction of our physical conceptions to explain away the results of the experiment I am going to propose; if they were not in accordance with the predictions of Euclidian geometry. Whitworth's method of manufacturing true planes, or 'surface plates,' depends solely on the objective axiom or hypothesis of congruence; and as the intersection of two such true planes is a straight line, it enables us to construct a series of material points occupying positions which form an 'unique group of the first order'—if we assume it to be possible to catalogue the positions of material points by the method of unique boundaries at all. Suppose then that I construct three such material 'straight edges,' each one yard long, and arrange them to form a triangle ABC. I then, by the same method, construct another straight edge, say one foot long, and by it measure off AD, AE along AB and AC, each equal to one foot, and place my one-foot rule with one end at D, and try whether I can at the same time place the other end, E', at E. No *a priori* reasoning can predict with certainty whether E' will fall upon E, or whether it will fall short, or over. The hypothesis of congruence, if objectively valid, tells me that whichever way it falls in one part of the universe it will fall the same way in any other, or at any other time. If, therefore, it fell short in one place and over in another, or short one day and over the next, we might indeed try to explain this by altering physical hypotheses, but it might conceivably be so difficult to frame a physical hypothesis to account for all the experiments, that we should in the end abandon the axiom, '*Es existiren in sich feste Körper*,' as objectively valueless. But meta-geometers draw a yet more far-reaching conclusion from this axiom. They can, namely, calculate the exact amount by which E' will fall short of, or over, E in the case of triangles of any linear dimensions whatever, if they are given the amount it falls short, or over, in one particular case. This is a perfectly legitimate deduction from the axiom, or hypothesis, of congruence, and it shows that we have a yet more searching test of the validity of this hypothesis, in that from the measurement of one particular case we could predict the measurement in any other. The objective application of the axiom of congruence is therefore empirical—in exactly the same sense, if not in quite the same degree, as the 'law of gravitation' is. And this is where, I believe, that Mr. Russell's idealistic bias has misled him.

He calls the axioms of projective geometry 'wholly *a priori*,' by which he means something more than that they are necessary merely in order to establish the conclusions of a particular science. He says that they *must* be true of 'any form of externality,' and that 'without them experience would be impossible.' Whatever qualifying or minimising explanations may be found in the rest of the book I am sure his readers will take this to mean that, in his view, we may apply *either* Euclidian *or* meta-geometrical calculations to practical space-measurements, with something more than empirical confidence. Indeed, he says distinctly that the empirical element in geometry "arises out of the alternatives of Euclidian and non-Euclidian space"; by which of course he understands Euclidian, spherical, or hyperbolic space. Now I have shown how these alternatives may be tested, not only as against each other, but as against non-congruent geometries, without any reference to light-theories. The result of one such test can not, even according to Mr. Russell, be predicted on *a priori* grounds. But the test once made with sufficient accuracy, the empirical element, according to Mr. Russell, is decided once for all. We could predict from one experiment exactly how far E' would fall short of, or over, E, in the case of any triangles, whenever or wherever measured; the only possibility of failure of our prediction lying in the possibilities of inaccuracy in our first experiment. Does Mr. Russell really believe that it is, on *a priori* grounds, *impossible* that we should have such an experience as that, say, in England E' should always fall short of E, and in Germany always fall over? Or is it *impossible* that the distance EE' should be any other function of the linear magnitudes of the triangles than those assigned by spherical or hyperbolic geometry?

There is of course one way by which Mr. Russell, or any idealist, might seek to evade this question. He might say that as a philosopher he had nothing to do with actual physical measurements, that he recognised no reality beyond subjective reality, and never intended his theories to have what I call objective validity. But, if so, Mr. Russell has failed to achieve the object he set himself in the Introduction to his book. He is still on the horns of the dilemma that 'none but a madman would throw a doubt on the validity of geometry, and none but a fool would deny its objective reference.' He has still to explain why practical men rely on the validity of Euclidian geometry, and, if he thinks them fools for doing so, why it is that predictions made on the strength of it always turn out true, within the limits of experimental error.

This is the test by which I proposed at the commencement of this paper to distinguish between realistic and idealistic theories of philosophy. It may be that the distinction is in reality only verbal: Mr. Muirhead, for example, when I read my paper on Natural Realism before this Society, called me an idealist, and I called him a realist in reply. It may even be that when idealistic philosophers talk about the absolute they really mean something, which, if only I could learn their language, I should recognise as something I have long been familiar with. But even modern philosophers are ordinary human beings in private life, and if this is the explanation of their theories they must know enough plain English to be able to translate them into ordinary language. There are of course plenty of philosophers who do use plain language, such as Locke, Mill, or even Herbert Spencer. And if I do not agree with their views it is easy to trace where, in my opinion, they have gone wrong. Though I may hold their theories to be erroneous, they are perfectly intelligible. But with Idealism in its modern form this is not the case. Modern philosophers talk the language of genuine Idealism; but instead of frankly abandoning its errors they alter the meanings of its terms, much, or little, according to the force of the criticisms directed against them. The consequence is that to attack them is like punching a bolster; you may make a dent, but soon after your back is turned it will bulge out to its original shape again. If anyone here can follow me in my theory, and agree in the conclusions I have reached, I shall hail him with pleasure as a natural realist; only in that case I must really beg him finally to sever any connection he may have had with *a priori* philosophy. If on the other hand he thinks there is a logical necessity which makes it *impossible* that triangles should have an excess in England and a defect in Germany—then I shall know what to think of him, as a philosopher.

EDWARD T. DIXON.

CAMBRIDGE,
       August, 1897.

www.ingramcontent.com/pod-product-compliance
Lightning Source LLC
Chambersburg PA
CBHW021957190326
41519CB00009B/1305

9783337008833